CLIMATE CHANGE AND THE FUTURE OF SUSTAINABILITY

The Impact on Renewable Resources

CLIMATE CHANGE AND THE FUTURE OF SUSTAINABILITY

The Impact on Renewable Resources

Edited by

Muyiwa Adaramola, PhD

APPLE ACADEMIC PRESS

Apple Academic Press Inc.
3333 Mistwell Crescent
Oakville, ON L6L 0A2
Canada

Apple Academic Press Inc.
9 Spinnaker Way
Waretown, NJ 08758
USA

Exclusive worldwide distribution by CRC Press, a member of Taylor & Francis Group

Printed in the United States of America on acid-free paper

International Standard Book Number-13: 978-1-77188-431-0 (Hardcover)

International Standard Book Number-13: 978-1-77188-432-7 (eBook)

Library and Archives Canada Cataloguing in Publication

Climate change and the future of sustainability : the impact on renewable resources / edited by Muyiwa Adaramola, PhD.

Includes bibliographical references and index.
Issued in print and electronic formats.
ISBN 978-1-77188-431-0 (hardcover).--ISBN 978-1-77188-432-7 (pdf)
1. Renewable energy sources--Forecasting. 2. Climatic changes--Forecasting. I. Adaramola, Muyiwa, editor

TJ808 C65 2016 333.79'4 C2016-901371-5 C2016-901372-3

Library of Congress Cataloging-in-Publication Data

Names: Adaramola, Muyiwa, editor.
Title: Climate change and the future of sustainability : the impact on renewable resources / Muyiwa Adaramola, PhD, editor.
Description: Toronto : Apple Academic Press, 2016. |
Includes bibliographical references and index.
Identifiers: LCCN 2016009030 (print) | LCCN 2016018084 (ebook) | ISBN 9781771884310 (hardcover : alk. paper) | ISBN 781771884327 ()
Subjects: LCSH: Renewable energy sources. | Renewable natural resources. | Climatic changes.
Classification: LCC TJ808 .C625 2016 (print) | LCC TJ808 (ebook) | DDC 333.79--dc23
LC record available at https://lccn.loc.gov/2016009030

Apple Academic Press also publishes its books in a variety of electronic formats. Some content that appears in print may not be available in electronic format. For information about Apple Academic Press products, visit our website at **www.appleacademicpress.com** and the CRC Press website at **www.crcpress.com**

About the Editor

MUYIWA S. ADARAMOLA, PhD

Dr. Muyiwa S. Adaramola earned his BSc and MSc in Mechanical Engineering from the Obafemi Awolowo University, Nigeria, and the University of Ibadan, Nigeria, respectively. He received his PhD in Environmental Engineering at the University of Saskatchewan in Saskatoon, Canada. He has worked as lecturer at the Obafemi Awolowo University and as a researcher at the Norwegian University of Science and Technology, Trondheim, Norway. Currently, Dr. Adaramola is a Professor in Renewable Energy at the Norwegian University of Life Sciences, Ås, Norway.

Contents

Acknowledgment and How to Cite

The editor and publisher thank each of the authors who contributed to this book. The chapters in this book were previously published elsewhere. To cite the work contained in this book and to view the individual permissions, please refer to the citation at the beginning of each chapter. Each chapter was carefully selected by the editor; the result is a book that looks at the impact of renewable resources on climate change from a variety of perspectives. The chapters included are broken into five sections, which describe the following topics:

- In chapter 1, Kulkarni and Huang deduce the centennial trends in the surface wind speed over North America from global climate model simulations and project that the current estimate of wind power potential for North America based on present-day climatology will not be significantly changed by the greenhouse gas forcing in the coming decades.
- In chapter 2, Wang and Prinn conclude that the intermittency of wind power on daily, monthly and longer time scales as computed in their simulations and inferred from their meteorological observations, poses a demand for one or more options to ensure reliability, including backup generation capacity, very long distance power transmission lines, and on-site energy storage, although each will have economic and/or technological challenges.
- In chapter 3, Saenz and Huang quantify the extent to which changes in atmospheric processes affect the downward solar radiation at the surface, as directly calculated by their climate models.
- In chapter 4, Panagea and her colleagues examine the relative contributions of temperature and irradiance, and predict a significant reduction due to the temperature increase, which will be, however, outweighed by the irradiance increase, resulting in an overall increase in photovoltaic systems.
- In chapter 5, Hamududu and Killingtveit evaluate the changes in global hydropower generation resulting from predicted changes in climate.
- In chapter 6, Dr. van Vliet and her colleagues examine the impact of climate change on water availability, and water temperature on electricity production and prices, using simulations of daily river flows and water temperatures under future climate to show declines in both thermoelectric and hydropower generating potential.

- In chapter 7, Dr. Koch and her colleagues calculate the hydroelectric power generation on a high temporal resolution of one hour for single hydropower plants within a large scale catchment, concluding that all hydropower plants of their studied watershed can be considered in a parallel way.
- In chapter 8, Dr. Rybach presents global statistical data on the current status of deep geothermal resource utilization for electricity generation.
- In chapter 9, Dr. Cocchi and her colleagues develop a model in order to simulate an air conditioning system with geothermal heat pump.
- In chapter 10, Dr. Garmsiri and colleagues compare sewer waste heat recovery with heat pumps using geothermal energy storage systems for a small community shared water heating system including commercial and institutional buildings, finding that the sewer heat exchanger method is relatively economical, since it has the smallest rate of return on investment for the selected community size.
- In chapter 11, Dr. Jwo and colleagues propose a new technique to directly adopt the wind force to drive heat pump systems, which could effectively reduce the energy conversion losses during the processes of wind force energy converting to electric energy and electric energy converting to kinetic energy.
- In chapter 12, Tiong and colleagues focus on the capability of renewable wind and solar energy in generating power for offshore application.
- In chapter 13, Kim and colleagues present an efficient plan for the application of a geothermal energy facility at the building structure planning phase.

List of Contributors

Florian Appel
VISTA GmbH, Remote Sensing in Geosciences, Gabelsbergerstrasse 51, Munich 80333, Germany

Heike Bach
VISTA GmbH, Remote Sensing in Geosciences, Gabelsbergerstrasse 51, Munich 80333, Germany

Sonia Castellucci
CIRDER, Università degli Studi della Tuscia, via San Camillo de Lellis, 01100 Viterbo, Italy

Yen-Lin Chen
Department of Computer Science and Information Engineering, National Taipei University of Technology, Taipei 10608, Taiwan

Chao-Chun Chien
Department of Energy and Refrigerating Air-Conditioning Engineering, National Taipei University of Technology, Taipei 10608, Taiwan

Zi-Jie Chien
Department of Energy and Refrigerating Air-Conditioning Engineering, National Taipei University of Technology, Taipei 10608, Taiwan

Silvia Cocchi
DAFNE, Università degli Studi della Tuscia, via San Camillo de Lellis, 01100 Viterbo, Italy

S. S. Dol
Department of Mechanical Engineering, Curtin University Sarawak Campus CDT 250, 98009 Miri, Sarawak, Malaysia

Shahryar Garmsiri
University of Ontario Institute of Technology, Oshawa, Ontario, Canada

Manolis G. Grillakis
Department of Environmental Engineering, Technical University of Crete, GR73100 Chania, Greece

Huei-Ping Huang
School for Engineering of Matter, Transport, and Energy, Arizona State University, Tempe, AZ 85281, USA

Byman Hamududu
Department of Hydraulic and Environment Engineering, Faculty of Engineering Science and Technology, Norwegian University of Science and Technology, Trondheim 7491, Norway

Young Jun Jang
Department of Plant/Architectural Engineering, Kyonggi University, Suwon-si, Gyeonggi-do 443-760; Korea

Ching-Song Jwo
Department of Energy and Refrigerating Air-Conditioning Engineering, National Taipei University of Technology, Taipei 10608, Taiwan

Aanund Killingtveit
Department of Hydraulic and Environment Engineering, Faculty of Engineering Science and Technology, Norwegian University of Science and Technology, Trondheim 7491, Norway

Gwang-Hee Kim
Department of Plant/Architectural Engineering, Kyonggi University, Suwon-si, Gyeonggi-do 443-760; Korea

Sangyong Kim
School of Construction Management and Engineering, University of Reading, Whiteknights, P.O. Box 219, Reading RG6 6AW, UK

Franziska Koch
Department of Geography, Ludwig-Maximilians-Universität München, Luisenstr. 37, Munich 80333, Germany

Seama Kouhi
University of Ontario Institute of Technology, Oshawa, Ontario, Canada

Aristeidis G. Koutroulis
Department of Environmental Engineering, Technical University of Crete, GR73100 Chania, Greece

Sujay Kulkarni
School for Engineering of Matter, Transport, and Energy, Arizona State University, Tempe, AZ 85281, USA

Wolfram Mauser
Department of Geography, Ludwig-Maximilians-Universität München, Luisenstr. 37, Munich 80333, Germany

Ioanna S. Panagea
Department of Environmental Engineering, Technical University of Crete, GR73100 Chania, Greece

Monika Prasch
Department of Geography, Ludwig-Maximilians-Universität München, Luisenstr. 37, Munich 80333, Germany

R. G. Prinn
Center for Global Change Science and Joint Program of the Science and Policy of Global Change, Massachusetts Institute of Technology, Cambridge, MA 02139, USA

Marc A. Rosen
University of Ontario Institute of Technology, Oshawa, Ontario, Canada

Dirk Rübbelke
Basque Centre for Climate Change, Alameda Urquijo, 4-4, E-48008 Bilbao, Spain and IKERBASQUE, Basque Foundation for Science, E-48011, Bilbao, Spain

Ladislaus Rybach
Institute of Geophysics, ETH Zurich, Sonneggstrasse 5, CH-8092 Zurich, Switzerland

Gerardo Andres Saenz
School for Engineering of Matter, Transport and Energy, Arizona State University, Tempe, AZ 85281, USA

Yoonseok Shin
Department of Plant/Architectural Engineering, Kyonggi University, Suwon-si, Gyeonggi-do 443-760; Korea

Y. K. Tiong
Department of Mechanical Engineering, Curtin University Sarawak Campus CDT 250, 98009 Miri, Sarawak, Malaysia

Ioannis K. Tsanis
Department of Environmental Engineering, Technical University of Crete, GR73100 Chania, Greece and Department of Civil Engineering, McMaster University, Hamilton, ON, Canada L8S 4L7

Andrea Tucci
DAFNE, Università degli Studi della Tuscia, via San Camillo de Lellis, 01100 Viterbo, Italy

Michelle T. H. van Vliet
Earth System Science—Climate Change and Adaptive Land and Water Management, Wageningen University and Research Centre, PO Box 47, 6700 AA Wageningen, The Netherlands and International Institute for Applied Systems Analysis, Schlossplatz 1, A-2361 Laxenburg, Austria

Stefan Vögele
Forschungszentrum Jülich, Institute of Energy and Climate Research–System Analyses and Technology Evaluation, D-52425 Jülich, Germany

C. Wang
Center for Global Change Science and Joint Program of the Science and Policy of Global Change, Massachusetts Institute of Technology, Cambridge, MA 02139, USA

Markus Weber
Commission for Geodesy and Glaciology, Bavarian Academy of Sciences and Humanities, Alfons-Goppel-Str. 11, Munich 80539, Germany

S. F. Wong
Department of Mechanical Engineering, Curtin University Sarawak Campus CDT 250, 98009 Miri, Sarawak, Malaysia

M. A. Zahari
Department of Mechanical Engineering, Curtin University Sarawak Campus CDT 250, 98009 Miri, Sarawak, Malaysia

Introduction

Renewable resources such as wind, solar, and geothermal are often perceived as being the answer to the fossil fuel crisis. Ironically, however, climate change may also negatively impact on these energy sources.

In a world where reducing carbon-based fuels usage will most likely be the norm, wind power is seen as valuable, in part, due to its lack of greenhouse gas emissions. However, climate change also presents risks and threats to wind resources. The major risks for wind power fall into two basic categories, which are changes in the wind resource distribution and risks to infrastructure.

All forms of renewable energy are somewhat sensitive to climate variation. While not as vulnerable to climate change as hydropower, wind resources will likely face some shifts in location, intensity, interval, and duration. However, those shifts are difficult to predict. While climate models do include air-circulation patterns, the models' outputs focus primarily on projected changes in temperature and precipitation, not wind flows, indicating a need for ongoing research in this area. Increased cloud cover due to changed climate patterns could also impact the availability of solar energy. Changed weather patterns will more certainly affect river levels, impacting hydropower plants.

What is certain is that today's models may not necessarily be accurate when predicting a world where climate change has become a reality. Because of this, the research collected in this compendium is of vital importance in order to begin creating a more accurate picture of what we can expect from renewable energy sources in the years ahead.

—Muyiwa Adaramola

The centennial trends in the surface wind speed over North America are deduced from global climate model simulations in the Climate Model Intercomparison Project—Phase 5 (CMIP5) archive. In Chapter 1, Kulkarni and Huang use the 21st century simulations under the RCP 8.5 scenario of greenhouse gas emissions, and find 5–10 percent increases per century in the 10 m wind speed over Central and East-Central United States, the Californian Coast, and the South and East Coasts of the USA in winter. In summer, climate models projected decreases in the wind speed ranging from 5 to 10 percent per century over the same coastal regions. These projected changes in the surface wind speed are moderate and imply that the current estimate of wind power potential for North America based on present-day climatology will not be significantly changed by the greenhouse gas forcing in the coming decades.

Meeting future world energy needs while addressing climate change requires large-scale deployment of low or zero greenhouse gas (GHG) emission technologies such as wind energy. The widespread availability of wind power has fueled substantial interest in this renewable energy source as one of the needed technologies. For very large-scale utilization of this resource, there are however potential environmental impacts, and also problems arising from its inherent intermittency, in addition to the present need to lower unit costs. To explore some of these issues, Wang and Prinn in Chapter 2 use a threedimensional climate model to simulate the potential climate effects associated with installation of wind-powered generators over vast areas of land or coastal ocean. Using wind turbines to meet 10% or more of global energy demand in 2100, could cause surface warming exceeding 1 ◦C over land installations. In contrast, surface cooling exceeding 1 ◦C is computed over ocean installations, but the validity of simulating the impacts of wind turbines by simply increasing the ocean surface drag needs further study. Significant warming or cooling remote from both the land and ocean installations, and alterations of the global distributions of rainfall and clouds also occur. These results are influenced by the competing effects of increases in roughness and decreases in wind speed on near-surface turbulent heat fluxes, the differing nature of land and ocean surface friction, and the dimensions of the installations parallel and perpendicular to the prevailing winds. These results are also dependent on the accuracy of the model used, and the realism of the methods applied to simulate wind turbines. Additional theory and new field observations will be required for their ultimate validation. Intermittency of wind power on daily, monthly and longer time scales as computed in these simulations and inferred from meteorological observations, poses a demand for one or more options to ensure reliability, including backup generation capacity, very long distance power transmission lines, and onsite energy storage, each with specific economic and/or technological challenges.

The projected changes in the downward solar radiation at the surface over North America for late 21st century are deduced from global climate model simulations with greenhouse-gas (GHG) forcing. A robust trend is found in winter over the United States, which exhibits a simple pattern of a decrease of sunlight over Northern USA. and an increase of sunlight over Southern USA. In Chapter 3, Saenz and Huang identify this structure in both the seasonal mean and the mean climatology at different times of the day. It is broadly consistent with the known poleward shift of storm tracks in winter in climate model simulations with GHG forcing. The centennial trend of the downward shortwave radiation at the surface in Northern USA. is on the order of 10% of the climatological value for the January monthly mean, and slightly over 10% at the time when it is midday in the United States. This indicates a nonnegligible influence of the GHG forcing on solar energy in the long term. Nevertheless, when dividing the 10% by a century, in the near term, the impact of the GHG forcing is relatively minor such that the estimate of solar power potential using present-day climatology will remain useful in the coming decades.

Solar power is the third major renewable energy, constituting an increasingly important component of global future—low carbon—energy portfolio. Accurate climate information is essential for the conditions of solar energy production, maximization, and stable regulation and planning. Climate change impacts on energy output projections are thus of crucial importance. In Chapter 4, by Panagea and colleagues, the effect of projected changes in irradiance and temperature on the performance of photovoltaic systems in Greece is examined. Climate projections were obtained from 5 regional climate models (RCMs) under the A1B emissions scenario, for two future periods. The RCM data present systematic errors against observed values, resulting in the need of bias adjustment. The projected change in photovoltaic energy output was then estimated, considering changes in temperature and insolation. The spatiotemporal analysis indicates significant increase in mean annual temperature (up to 3.5°C) and mean total radiation (up to 5 W/m^2) by 2100. The performance of photovoltaic systems exhibits a negative linear dependence on the projected temperature increase which is outweighed by the expected increase of total radiation resulting in an up to 4% increase in energy output.

Currently, hydropower accounts for close to 16% of the world's total power supply and is the world's most dominant (86%) source of renewable electrical energy. The key resource for hydropower generation is runoff, which is dependent on precipitation. The future global climate is uncertain and thus poses some risk for the hydropower generation sector. The crucial question and challenge then is what will be the impact of climate change on global hydropower generation and what are the resulting regional variations in hydropower generation potential? Chapter 5, by Hamududu and Killingtveit, is a study that aims to evaluate the changes in global hydropower generation resulting from predicted changes in climate. The study uses an ensemble of simulations of regional patterns of changes in runoff, computed from global circulation models (GCM) simulations with 12 different models. Based on these runoff changes, hydropower generation is estimated by relating the runoff changes to hydropower generation potential through geographical information system (GIS), based on 2005 hydropower generation. Hydropower data obtained from EIA (energy generation), national sites, FAO (water resources) and UNEP were used in the analysis. The countries/states were used as computational units to reduce the complexities of the analysis. The results indicate that there are large variations of changes (increases/decreases) in hydropower generation across regions and even within regions. Globally, hydropower generation is predicted to change very little by the year 2050 for the hydropower system in operation today. This change amounts to an increase of less than 1% of the current (2005) generation level although it is necessary to carry out basin level detailed assessment for local impacts which may differ from the country based values. There are many regions where runoff and hydropower generation will increase due to increasing precipitation, but also many regions where there will be a decrease. Based on this evaluation, it has been concluded that even if individual countries and regions may experience significant impacts, climate change will not lead to significant changes in the global hydropower generation, at least for the existing hydropower system.

Recent warm, dry summers showed the vulnerability of the European power sector to low water availability and high river temperatures. Climate change is likely to impact electricity supply, in terms of both water availability for hydropower generation and cooling water usage for thermoelectric power production. In Chapter 6, van Vliet and colleagues show the impacts of climate change and changes in water availability and water temperature on European electricity production and prices. Using simulations of daily river flows and water temperatures under future climate (2031–2060) in power production models, we show declines in both thermoelectric and hydropower generating potential for most parts of Europe, except for the most northern countries. Based on changes in power production potentials, we assess the cost-optimal use of power plants for each European country by taking electricity import and export constraints into account. Higher wholesale prices are projected on a mean annual basis for most European countries (except for Sweden and Norway), with strongest increases for Slovenia (12–15%), Bulgaria (21–23%) and Romania (31–32% for 2031–2060), where limitations in water availability mainly affect power plants with low production costs. Considering the long design life of power plant infrastructures, short-term adaptation strategies are highly recommended to prevent undesired distributional and allocative effects.

Climate change has a large impact on water resources and thus on hydropower. Hydroelectric power generation is closely linked to the regional hydrological situation of a watershed and reacts sensitively to changes in water quantity and seasonality. In Chapter 7, Koch and colleagues modelled the development of hydroelectric power generation in the Upper Danube basin for two future decades, namely 2021–2030 and 2051–2060, using a special hydropower module coupled with the physically-based hydrological model PROMET. To cover a possible range of uncertainties, 16 climate scenarios were taken as meteorological drivers which were defined from different ensemble outputs of a stochastic climate generator, based on the IPCC-SRES-A1B emission scenario and four regional climate trends. Depending on the trends, the results show a slight to severe decline in hydroelectric power generation. Whilst the mean summer values indicate a decrease, the mean winter values display an increase.

To show past and future regional differences within the Upper Danube basin, three hydropower plants at individual locations were selected. Interannual differences originate predominately from unequal contributions of the runoff compartments rain, snow- and ice-melt.

In Chapter 8, Rybach presents the current status of deep geothermal resource utilization for electricity generation based on global statistical data. Particular attention is paid to growth rates. The rates are compared with those of other renewable energies (biomass, hydro, solar photovoltaic (PV), wind). Whereas wind and solar PV exhibit annual growth rates of 25%–30% since 2004, geothermal growth is only about 5% per year. Geothermal electricity production (in TW·h/yr) was higher until 2011 than from solar PV, but is now clearly falling behind. So far the global geothermal electricity generation is provided nearly entirely by hydrothermal resources, which exist only under specific geologic conditions. Further development (=increasing production capacity) based on this resource type alone will therefore hardly accelerate to two-digit (>10% per year) growth rates. Faster growth can only be achieved by using the ubiquitous petrothermal resources, provided that the key problem will be solved: establishing a universally applicable technology. This would enable to create, at any requested site, feasible and efficient deep heat exchangers for enhanced geothermal systems (EGS) power plants—irrespective of the local subsurface conditions. Goals and challenges of this technology are addressed.

The need to address climate change caused by greenhouse gas emissions attaches great importance to research aimed at using renewable energy. Geothermal energy is an interesting alternative concerning the production of energy for air conditioning of buildings (heating and cooling), through the use of geothermal heat pumps. In Chapter 9, by Cocchi and colleagues, a model has been developed in order to simulate an air conditioning system with geothermal heat pump. A ground source heat pump (GSHP) uses the shallow ground as a source of heat, thus taking advantage of its seasonally moderate temperatures. GSHP must be coupled with geothermal exchangers. The model leads to design optimization of geothermal heat exchangers and to verify the operation of the geothermal plant.

The consumption of hot water represents a significant portion of national energy consumption and contributes to concerns associated with global climate change. Utilizing heat recovered from the sewer, or the stored heat by utilizing heat pumps with a borehole geothermal energy storage system, are simple and effective ways of heating water for domestic purposes. Reclaiming heat from the waste warm water that is discharged to the sewer or stored heat in a borehole geothermal energy storage system can help reduce natural gas energy consumption as well as the associated energy costs and greenhouse gas emissions. In Chapter 10, by Garmsiri and colleagues, sewer waste heat recovery is compared with heat pumps using geothermal energy storage systems for a small community shared water heating system including commercial and institutional buildings. It is found that the sewer heat exchanger method is relatively economical as it has the smallest rate of return on investment for the selected community size. The findings also demonstrate a reduction occurs in natural gas consumption and fewer CO_2 gas emissions are emitted to the atmosphere. The results are intended to allow energy technology suppliers to work with communities while accounting appropriately for economic issues and CO_2 emissions associated with these energy technologies.

The requirements of providing electric energy through the wind-forced generator to the heat pump for water cooling and hot water heating grow significantly by now. In Chapter 11, Jwo and colleagues propose a new technique to directly adopt the wind force to drive heat pump systems, which can effectively reduce the energy conversion losses during the processes of wind force energy converting to electric energy and electric energy converting to kinetic energy. The operation of heat pump system transfers between chiller and heat that are controlled by a four-way valve. The theoretical efficiency of the traditional method, whose heat pump is directly forced by wind, is 42.19%. The experimental results indicated average value for cool water producing efficiency of 54.38% in the outdoor temperature of 35°C and the indoor temperature of 25°C and the hot water producing efficiency of 52.25% in the outdoor temperature and the indoor temperature both of 10°C. The authors proposed a method which can improve the efficiency over 10% in both cooling and heating.

Renewable energy is an energy which is freely available in nature such as winds and solar energy. It plays a critical role in greening the energy sector as these sources of energy produce little or no pollution to environment. Chapter 12, by Tiong and colleagues, will focus on capability of renewable energy (wind and solar) in generating power for offshore application. Data of wind speeds and solar irradiation that are available around SHELL Sabah Water Platform for every 10 minutes, 24 hours a day, for a period of one year are provided by SHELL Sarawak Sdn. Bhd. The suitable wind turbine and photovoltaic panel that are able to give a high output and higher reliability during operation period are selected by using the tabulated data. The highest power output generated using single wind energy application is equal to 492 kW while for solar energy application is equal to 20 kW. Using the calculated data, the feasibility of renewable energy is then determined based on the platform energy demand.

Chapter 13, by Kim and colleagues, aims to present an efficient plan for the application of a geothermal energy facility at the building structure planning phase. Energy consumption, energy cost and the primary energy consumption of buildings were calculated to enable a comparison of buildings prior to the application of a geothermal energy facility. The capacity for energy savings and the costs related to the installation of such a facility were estimated. To obtain more reliable criteria for economic feasibility, the lifecycle cost (LCC) analysis incorporated maintenance costs (reflecting repair and replacement cycles based on construction work specifications of a new renewable energy facility) and initial construction costs (calculated based on design drawings for its practical installation). It is expected that the findings of this study will help in the selection of an economically viable geothermal energy facility at the building construction planning phase.

PART I

WIND ENERGY AND CLIMATE CHANGE

CHAPTER 1

Changes in Surface Wind Speed over North America from CMIP5 Model Projections and Implications for Wind Energy

SUJAY KULKARNI AND HUEI-PING HUANG

1.1 INTRODUCTION

The rapid technological developments in the past decade have established wind energy as one of the major alternatives to fossil-fuel based energy. The potential of wind power generation in the United States alone, including off-shore and on-shore capacity, is estimated to be about 15000 GW (e.g., Lopez et al. [1]). This estimate generally does not take into account future climate changes which may alter the pattern and strength of near-surface wind at desirable locations for wind farms (Freedman et al. [2], Ren [3]). Worldwide, long-term projections of decadal-to-centennial climate changes due to anthropogenic emission of greenhouse gases (GHG) have been systematically carried out by climate modeling groups that participate in the Climate Model Intercomparison Project-Phase 5 (CMIP5, Taylor et al. [4], cmip5-pcmdi.llnl.gov/cmip5), in close association with the Intergovernmental Panel on Climate Change (IPCC) of the United Nations (IPCC [5]). While climate model outputs from CMIP5 and its predecessors have been widely used to project regional changes in temperature and hydrological cycles (e.g., Seager et al. [6], Baker and Huang [7]), few studies have used the datasets to project future changes in surface wind. Notably, Pryor and Barthelmie [8] analyzed the regional model simulations in NARCAAP (Mearns et al. [9]), constrained by the global model projections from CMIP3 (Meehl et al. [10]), to conclude that GHG-induced climate change will not significantly affect wind power potential in the United States in the coming decades. As a contribution to this under-explored area of research, this study will use a subset of the newer CMIP5 model data to construct the GHG-induced trends in the near-surface wind speed over North America.

The horizontal resolution of the global climate models in CMIP5 is typically around 100–150 km in midlatitudes. It is understood that this is not fine enough to resolve detailed topography in the mesoscale and submesoscale, which can have nontrivial influences on the low-level wind field. Nevertheless, the information from the global models provide the first-order picture of the changes in the large-scale flow, which will form the basis for future efforts to downscale the global model output to regional and urban scales. The CMIP5 simulations for the 21st Century are driven by the radiative forcing deduced from different scenarios of anthropogenic emissions of GHG and industrial aerosols. Regional climate changes due to land-use changes (e.g., urbanization) or even the influence of large-scale wind farms (e.g., Keith et al. [11] and Adams and Keith [12]) are not covered by the 21st century scenarios in CMIP5 and are not considered in this work.

1.2 DATASETS

Five models, from CMIP5, EC-Earth, IPSL-CM5-LR, GISS-E2-H, CSIRO-MK 3.6.0, and ACCESS 1.0 (listed in Table 1), are used in this study. By first examining the scatter plots of the indices of large-scale wind fields (in the manner of Paek and Huang [13]) over the Pacific-North American sector, the five models were selected as a subset that at least reflects the diversity (in terms of model resolution and biases) of the over 30 models in CMIP5. For example, IPSL-CM5-LR and GISS-E2-H substantially underestimate and CSIRO-MK 3.6.0 overestimates the Low Level Jet over North America, while the other two models produce only small biases in that feature (not shown). For our purpose of deducing trends, the historical runs for the 20th Century and the corresponding 21st century runs under the representative concentration pathways (RCP) 8.5 scenario are used. As a brief background, the RCP8.5 scenario imposes 8.5 W/m^2 of radiative forcing, induced by the projected increase in GHG concentration, to the atmosphere towards the end of the 21st Century. It produces an increase in global mean surface air temperature which ranges from +2.6 to +4.8°C over the 21st Century from the projections by the majority of CMIP5 models (IPCC [5]).

The global models in CMIP5 typically have very few vertical levels within the planetary boundary layer. Given that wind turbines are usually at 80–100 m height, at which there is no direct model output, the closest standard output variables that we can use from CMIP5 are the surface wind speed and the vector wind field at 10 m height as calculated from boundary layer parameterization schemes. We will use the standard monthly mean archives of those variables from CMIP5. It is worth noting that, consistent with our purpose, the monthly mean of surface wind speed in the archive is the monthly average of the wind speed calculated at daily or subdaily frequency. While the wind speed at 10 m is generally less than that at 80–100 m height, the two are highly correlated and can be related by the Hellman exponent and wind gradient equation used for wind turbines (e.g., Kaltschmitt et al. [14]). Thus, we analyze the 10 m wind as a close proxy of the actual wind at the turbine height.

The simulations from the last two decades of the historical and RCP8.5 runs are used to deduce the trends. More precisely, the centennial trend is defined as the climatology of 2079–2099 minus the climatology of 1979–1999. Winter and summer will be analyzed separately. The 10 m wind data from the NCEP-DOE reanalysis-2 (Kanamitsu et al. [15], data obtained from the archive at http://www.esrl.noaa.gov/psd/) for 1979–1999 will also be used to cross validate the CMIP5 historical runs.

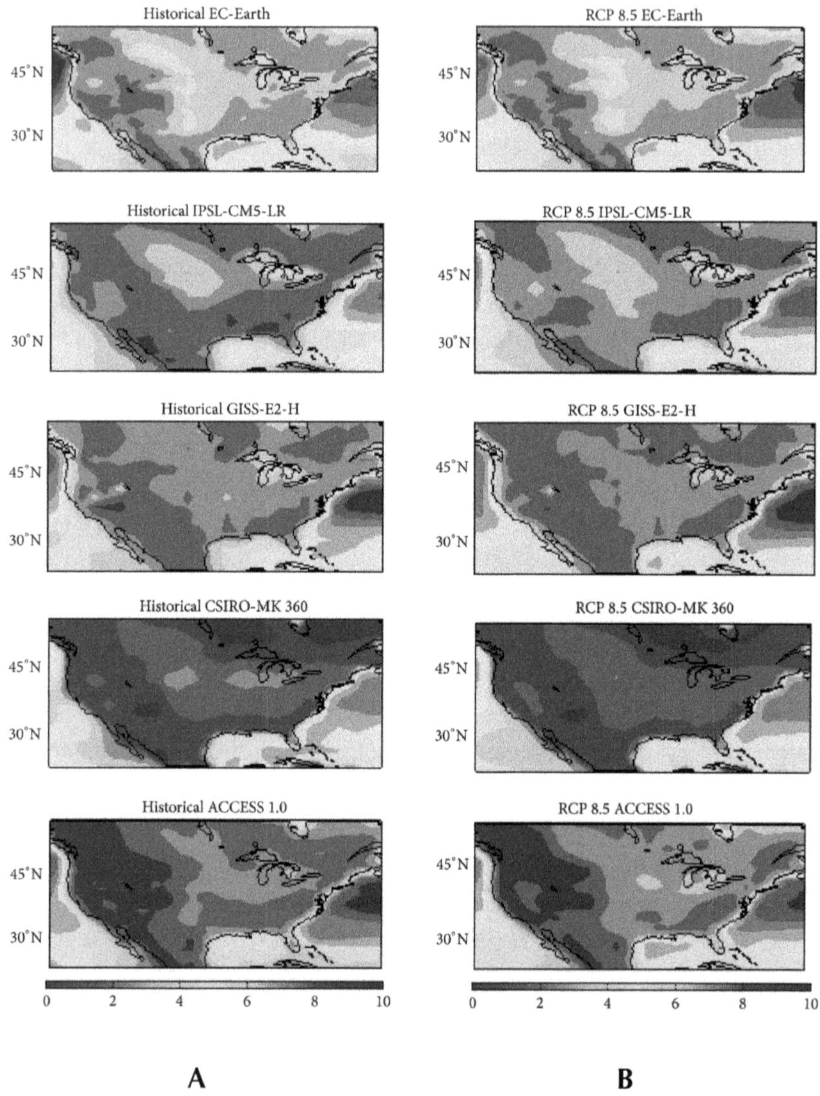

FIGURE 1: The climatology of surface wind speed over North America for winter (DJF) from the 20th century historical runs (a) and 21st century RCP 8.5 runs using five CMIP5 models as labeled at the top of each panel. The color scale, in m/s, is shown at bottom. Green and red colors represent lower and higher wind speed.

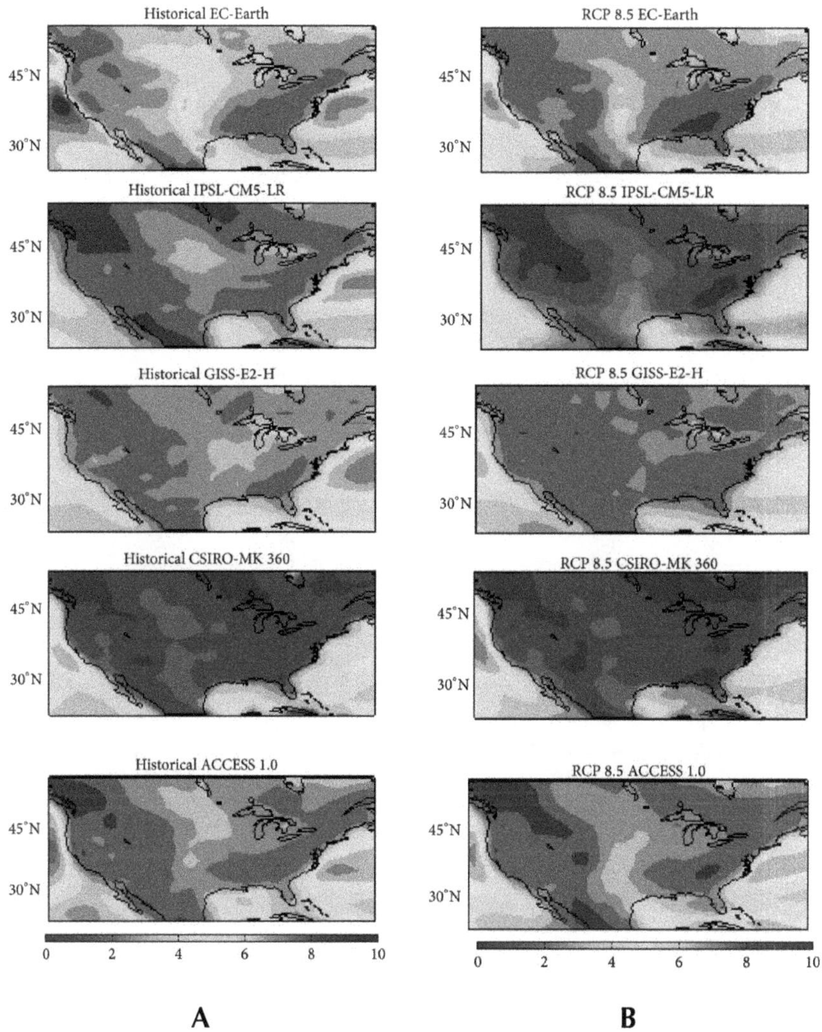

FIGURE 2: Similar to Figure 1 but for the climatology for summer (JJA). (a) and (b) are the historical and RCP 8.5 runs.

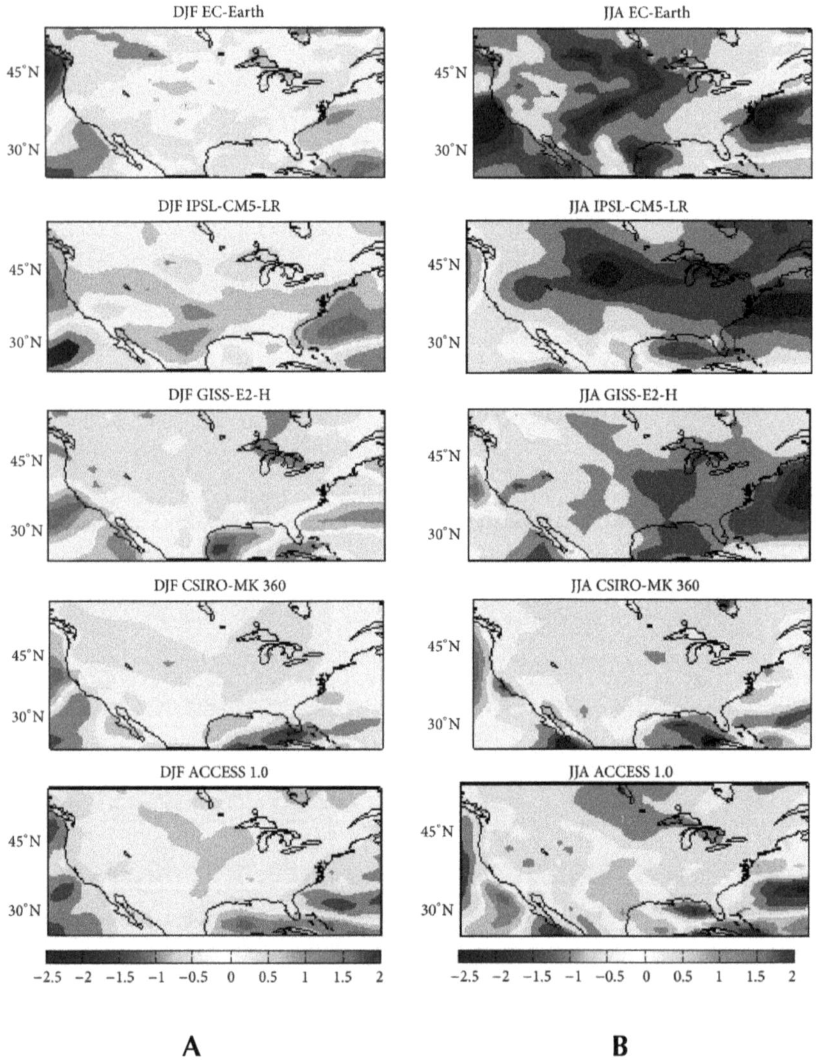

FIGURE 3: The trend [(2079–99) minus (1979–99)] in the surface wind speed over North America for winter (a) and summer (b), from five CMIP5 models as labeled at the top of each panel. The color scale, in m/s (per century), is shown at bottom. Blue and red indicate a decrease and an increase, respectively, in the surface wind speed.

1.3 SURFACE WIND SPEED IN PRESENT AND FUTURE CLIMATE

Figure 1 shows the climatology of the surface (10 m) wind speed over North America for the winter season (December–February) constructed from the last two decades of the 20th century historical runs (Figure 1(a)) and the last two decades of the 21st century RCP 8.5 runs (Figure 1(b)) using five different models in CMIP5. Figure 2 is similar to Figure 1 but for summer (June–August). For the 20th century simulations, the models produce the common first-order features with the highest wind speed over the oceans and relatively higher wind speed over the Great Plains compared to the Rockies and the Southeastern USA. The wind speed is higher in winter than in summer overall. These first-order features are also produced by the 21st century runs, giving the first indication that the GHG-induced climate change does not dramatically alter the surface wind field. Within either group of the 20th or 21st century runs, notable differences exist among the models. For example, in winter, GISS-E2-H and ACCESS 1.0 produce considerably stronger surface wind off the East Coast of the USA than other models; IPSL-CM5-LR and EC-Earth produce a more distinctive local maximum of surface wind over North-Central USA which is less visible in the simulations by the other three models. It is also interesting to note that only EC-Earth produces local surface wind maximum over the Great Lakes. This is because the model has the highest resolution among the five (see Table 1), high enough to partially resolve the lakes. The fine structures mentioned above that are unique to an individual model tend to exist in both the 20th and 21st century simulations by that model. This indicates that the model bias remains similar under the GHG forcing in the 21st century. In other words, if one defines the trend as the difference between the 21st century climatology and 20th century climatology, both from the same model, the bias would cancel itself. Thus, the trend so deduced can still be meaningful even if the model has biases.

Figure 3 shows the trends in the surface wind speed, defined as the 2079–2099 climatology minus the 1979–1999 climatology, over North America for winter (Figure 3(a)) and summer (Figure 3(b)), based on the simulations by the five models shown in Figures 1 and 2. The models produce diverse responses to GHG forcing. For example, IPSL-CM5-LR produces a positive trend in winter and negative trend in summer over almost the entire North American sector, while the responses in the CSIRO-MK 3.6.0 model are muted for both seasons. Nevertheless, when averaged across the models, the GHG-induced trends in the surface wind speed are overall an increase in winter and a decrease in summer over the North American continent. The increase in the surface wind speed in winter is broadly consistent with the enhancement of the eastward tropospheric jet stream aloft (which is a main feature in winter) found in previous analyses of the CMIP5 zonal wind data (Paek and Huang [13]).

TABLE 1: List of the CMIP5 models used in this study.

Model	Institution	Resolution
EC-Earth	EC-Earth consortium (multiple)	320 × 160/T159 (L62)
IPSL-CM5-LR	Institut Pierre-Simon Laplace (France)	96 × 96 (L39)
GISS-E2-H	NASA Goddard Institute for Space Studies (USA)	144 × 90 (L40)
CSIRO-MK 3.6.0	Commonwealth Scientific and Industrial Research Organisation (CSIRO) and Queensland Climate Change Centre of Excellence (Australia)	192 × 96/T63 (L31)
ACCESS 1.0	CSIRO and Bureau of Meteorology (Australia)	192 × 145 (L38)

The determination of the trends in Figure 3 is entirely based on models. As noted, if the model bias is not significantly affected by the GHG forcing in the 21st century, by taking the difference between the 21st and 20th century runs, the bias would cancel itself. This philosophy is also adopted by the IPCC in its assessment reports on future climate (IPCC [5]). Nevertheless, for completeness, we should compare selected models with the 20th century reanalysis to affirm that the biases are not excessive. Figure 4 shows the 1979–1999 climatology (averaged over all seasons) of the surface wind speed from NCEP-DOE reanalysis-2

(Figure 4(a)), along with its counterparts from the historical runs using GISS-E2-H (Figure 4(b)) and EC-Earth (Figure 4(c)). The overall patterns in reanalysis and model simulations are similar, although GISS-E2-H slightly underestimates the wind speed over West-Central US while EC-Earth overestimates it. A more complicated picture emerges if one further compares the climatology of the u- and v-components of the 10-meter wind. Figure 5(a) is similar to Figure 4 but for the v-component of surface wind and Figure 5(b) is for the u-component of it. Although EC-Earth has a larger bias in the surface wind speed, it simulates the v-component of the wind field better than GISS-E2-H. The bias in EC-Earth is mainly in the u-component. The two cases in Figures 4 and 5 suffice to illustrate that the model biases have somewhat complicated patterns but are not excessive in their magnitude. Also, a further examination did not reveal a simple correspondence between the pattern of the bias and the pattern of the trend.

1.4 REGIONAL SURFACE WIND FIELDS

With the changes in the surface wind speed shown in Figure 3, one may ask if there are also changes in the wind direction. The maps of the 10-meter wind fields, for selected models and regions with notable changes in wind speed, are shown in Figures 6 and 7. Figures 6(a) and 7(a) show the historical run and Figures 6(b) and 7(b) show the corresponding RCP 8.5 run. Wind fields are shown as the arrows, with the magnitude of the wind vector imposed in the background as the color shading. Figure 6 shows the EC-Earth simulations for Central USA (top) and the East Coast of the USA (bottom) for summer. Figure 7 shows the GISS-E2-H simulations for the Southern USA and part of Gulf of Mexico (top) and West Coast of the USA (bottom), both for winter. While significant changes in the wind direction are found in a few isolated places, for example, Illinois in the top row of Figure 6, and Pennsylvania and off the coast of New Jersey in the bottom row of Figure 6, for most regions shown in Figures 6 and 7 the GHG forcing does not induce major changes in the wind direction and the patterns of surface wind.

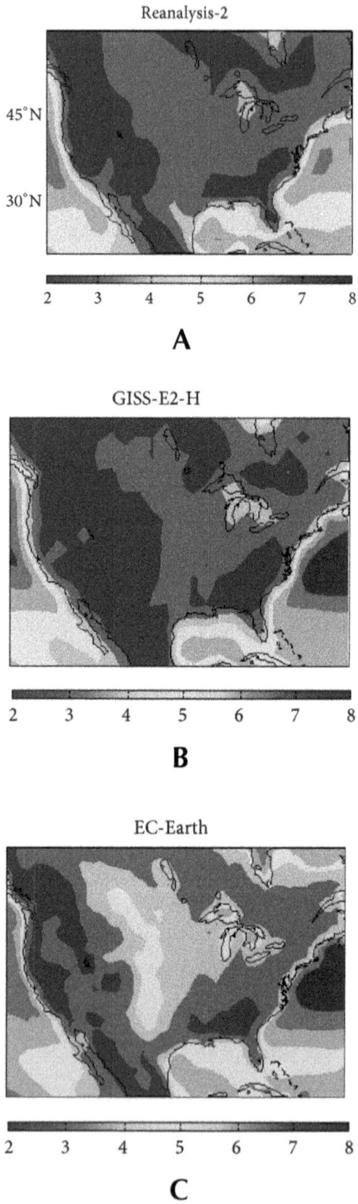

FIGURE 4: A comparison of the 1979–1999 climatology of the surface wind speed from NCEP-DOE reanalysis-2 (a) and the historical runs with two models (b) GISS-E2-H and (c) EC-Earth in CMIP5. The color scale, in m/s, is shown at bottom with red color indicating high wind speed.

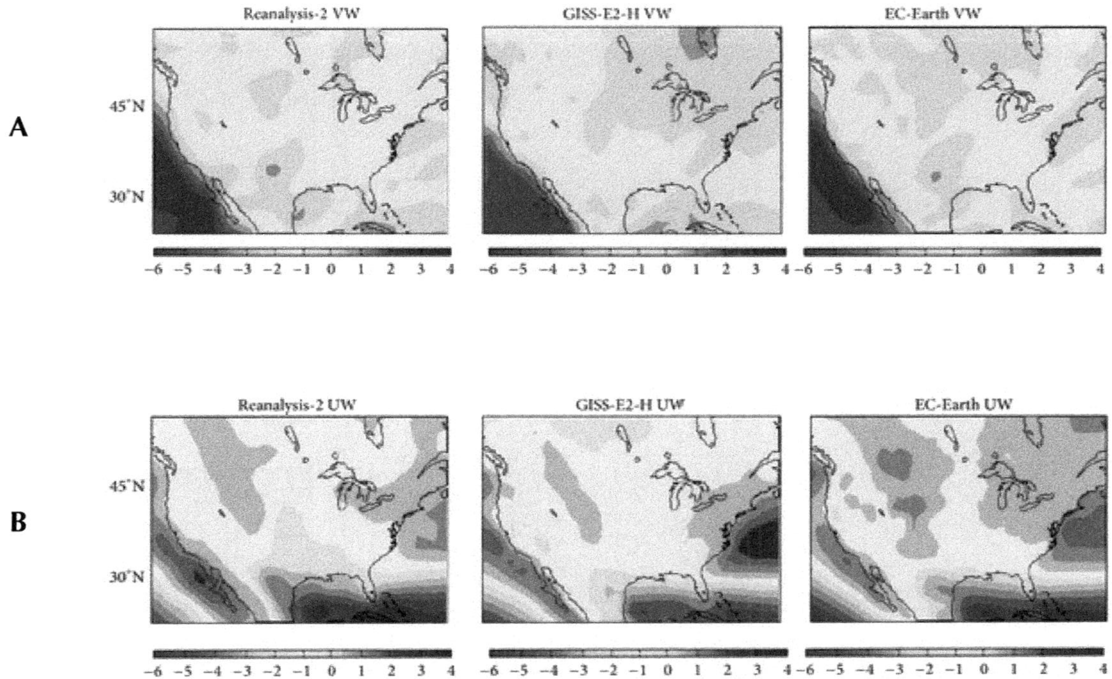

FIGURE 5: A comparison of the 1979–1999 climatology of the v-component (a) and u-component (b) of the 10 m wind over North America from reanalysis-2 (left), GISS-E2_H historical run (middle) and EC-Earth historical run (right). The color scale, in m/s, is shown at bottom. Red and green indicate positive and negative velocities.

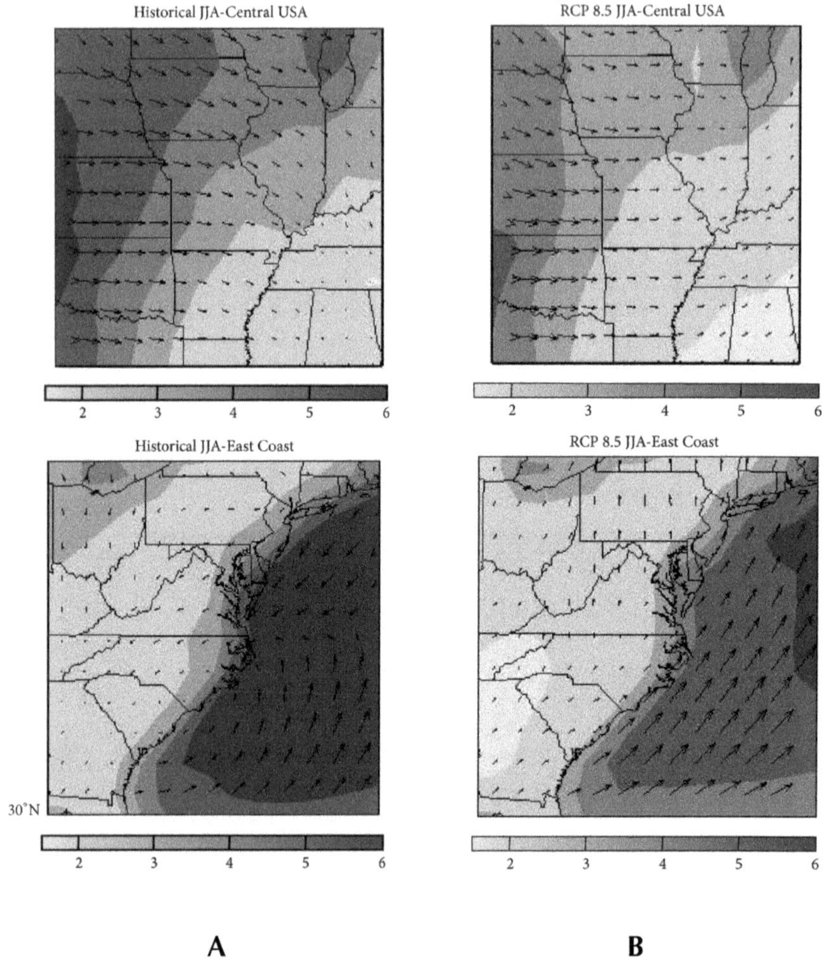

FIGURE 6: Detailed maps of the 10 m velocity fields for selected regions from the EC-Earth model simulations in CMIP5. (a) shows the 20th century historical runs and (b) the 21st century RCP 8.5 runs. Both are the average over the last two decades of the respective runs, and over summer (JJA) only. Top: Central United States. Bottom: the East Coast of the United States. The arrows indicate the climatological wind field and the color shading indicates the magnitude of the wind vectors shown. The color scale for the latter, in m/s, is shown at the bottom.

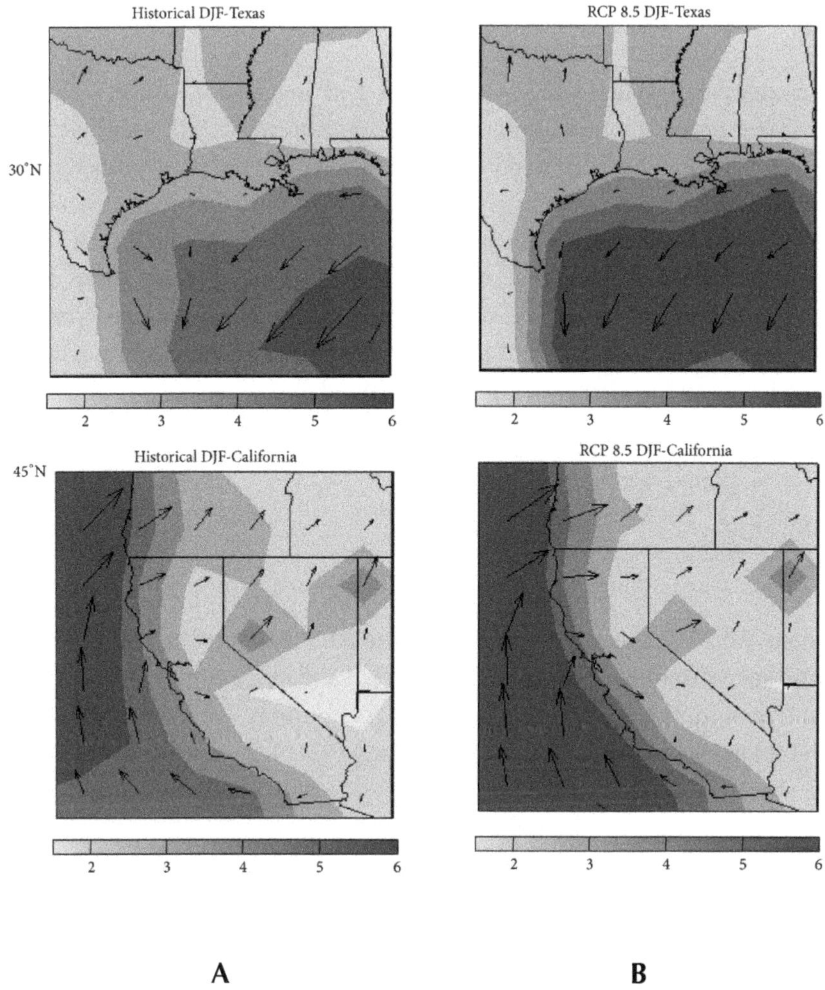

FIGURE 7: Similar to Figure 6 but for the surface wind fields from GISS-E2-H simulations for winter (DJF) and for two different regions. (a) and (b) are the historical and RCP 8.5 runs, respectively. Top: The Southern USA and part of Gulf of Mexico. Bottom: West Coast of the USA and off shore of California.

1.5 DISCUSSION

Our analysis has used the wind speed and horizontal velocity field at 10-meter height that are directly available from the CMIP5 archive. It is understood that the 10 m wind is used as a proxy of the wind at the turbine height of 80–100 m, which is typically stronger than the wind at near surface. Given so, a more useful measure of the influence of the GHG forcing is perhaps the percentage change, instead of the absolute value of the change, in the 10 m wind speed. At a grid point (i, j), where i and j are the indices for longitude and latitude, the multimodel average of the percentage change in the 10 m wind speed is defined as

$$\mu_{i,j} = \frac{1}{5}\sum_{k=1}^{5}\frac{(WS21)_{k,i,j} - (WS20)_{k,i,j}}{(WS20)_{k,i,j}} \tag{1}$$

where WS21 is the wind speed from the RCP 8.5 runs and WS20 is the wind speed from the historical runs and k is the index for the model. Since the five models have different horizontal resolutions, the CMIP5 data were first interpolated onto the same grid (using that of the reanalysis-2) before the statistics were calculated. The calculation of $\mu_{i,j}$ would not be meaningful over the regions where the surface wind speed (WS20) is very small, where wind turbines are also less likely to be built. To exclude those regions, we consider that most of the high capacity wind turbines operate above 5 m/s for practical energy production. By Hellman exponent and wind gradient equation used for wind turbines (e.g., Kaltschmitt et al. [14]), the wind speed at 80 m is typically 1.5 to 2 times that of the wind speed at 10 m height. Thus, we will neglect the regions with the 10 m wind speed less than 2 m/s. (If at least one model meets this criterion at a given grid point, that grid point is excluded from the calculation of $\mu_{i,j}$.) The maps of $\mu_{i,j}$ are shown for winter in Figure 8(a) and for summer in Figure 8(b). The white areas in Figure 8 are where either the climatology of the surface wind is small or the percentage change of the surface wind is small. The intramodel standard deviation (as a measure of the deviation from the multimodel mean, $\mu_{i,j}$) of the percentage change for the two seasons is also shown in Figure 9. The standard deviation is calculated only where the mean is calculated. For the convenience of plotting the result, in Figure 9, the standard deviation is set to zero over the areas where it is not calculated. In winter when the climatological surface wind is stronger overall, we find a moderate increase of 5–10% of the near surface wind speed over the Central and North-Central USA and the coastal regions in California and along the South and East Coasts of the USA. Using Betz's law (wind power proportional to the cube of wind speed), the equivalent changes in wind power potential would be approximately 15–30% per century over the colored areas in Figure 8(a). In summer, a decrease in wind speed at a similar range of 5–10% is found over the aforementioned coastal regions. A greater decrease, close to 20%, is found over isolated locations in West and West-Central USA. Nevertheless, those values are less reliable since they are associated with high intramodel standard deviation (compare the Figures 8(a) and 9(a)), indicating that the higher percentage of change is contributed by one or a small number of outliers.

In the preceding analysis we converted the GHG-induced change in the 10 m wind speed to an estimate of the change in wind power potential by simply applying the cubic law to the wind speed in the 20th and 21st century, then calculating the percentage change in "wind speed cubed." We used this simple approach because the CMIP5 archive does not provide the detailed wind and temperature profiles in the lower boundary layer. (Air temperature is available at 2 m only.) Note that the wind speed at the turbine height, U, is approximately related to the wind speed at a reference height (10 m in our case), U_R, by the relation of $(U/U_R) = (Z/Z_R)^{\alpha}$, where Z and Z_R are the heights of the turbine and the reference level and α (~0.14 for a neutrally stable profile) is an adjustable parameter (e.g., Peterson and Hennessey Jr. [16]). Thus, we obtained the estimate of the percentage change in wind power potential by implicitly assuming that α, or the static stability profile in the lower boundary layer, is not changed by the GHG forcing in the future. A validation of this assumption is beyond the scope of this study but will be a useful future work for climate modeling with high vertical resolutions.

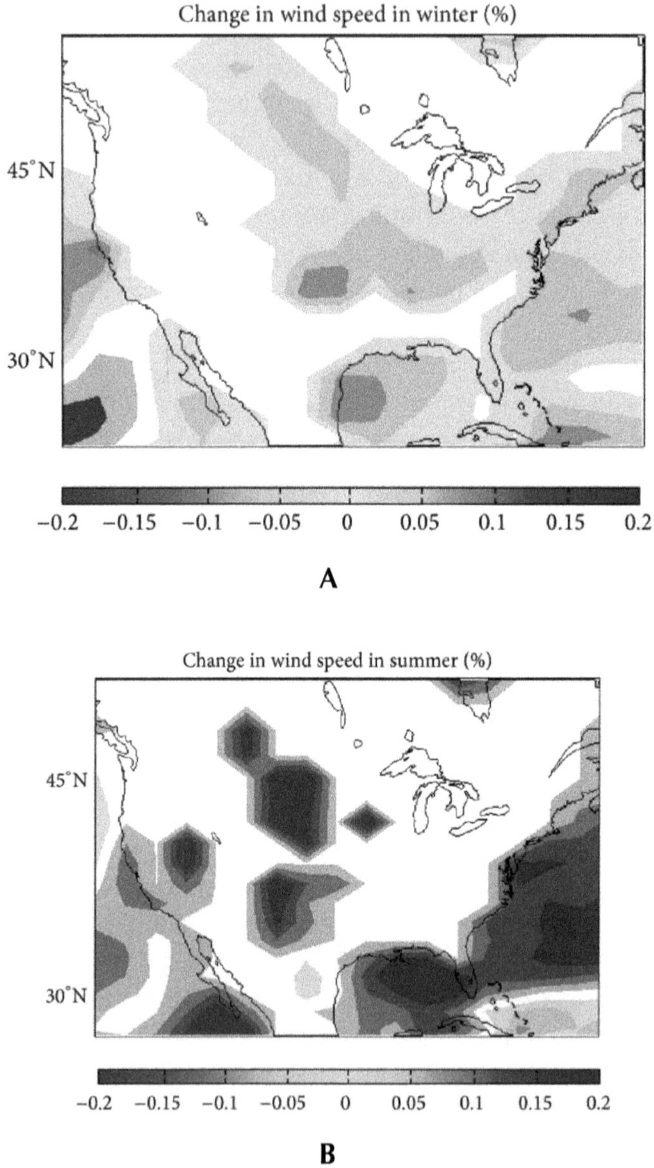

FIGURE 8: The multimodel mean of the percentage change in surface wind speed over North America for winter (a) and summer (b). See text for definition. Red indicates an increase and green indicates a decrease in wind speed. The values range from −20% to 20%, as indicated by the color scale at the bottom.

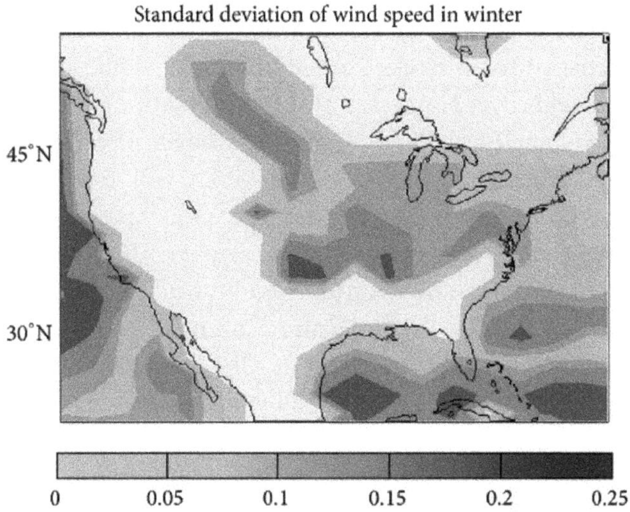

A

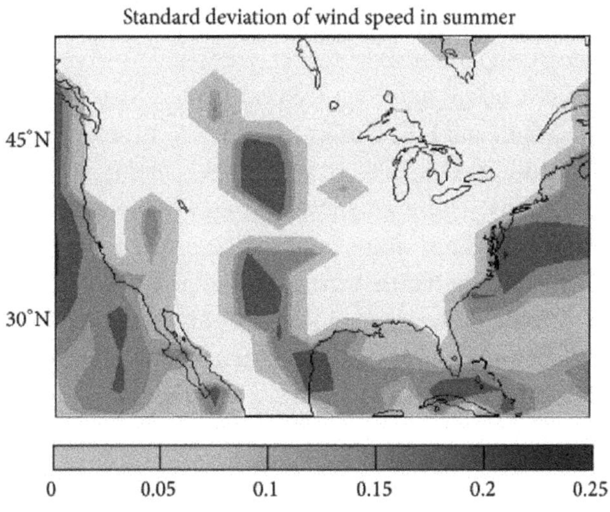

B

FIGURE 9: The intramodel standard deviation, with respect to the mean, as shown in Figure 8, of the percentage change in surface wind speed. (a) and (b) are for winter and summer, respectively. Where the standard deviation is not calculated (see text), it is set to zero and colored in gray. The color scale is shown at bottom.

Our results of the changes in surface wind speed and wind direction suggest that the GHG forcing (as used in CMIP5 simulations under the RCP 8.5 scenario) has a moderate, but not major, influence on the near-surface wind fields over North America. This broadly agrees with the conclusion of Pryor and Barthelmie [8] that the estimate of wind power potential over the USA using present-day climatology will remain useful in the coming decades. Note that the trend considered in this study is defined as the centennial change over the whole 21st century. The equivalent change over only the next 50 years (as discussed by Pryor and Barthelmie [8]) would be smaller. The RCP 8.5 scenario chosen for our analysis is among the ones with a higher estimate of future GHG emissions. If the RCP 4.5 scenario was chosen, the projected trend would also be smaller.

1.6 CONCLUDING REMARKS

Using 5 models from the CMIP5 archive and comparing the RCP 8.5 runs with historical runs, moderate centennial trends in the 10 m wind speed are projected over North America. In winter, we found 5–10 percent increases per century over Central and East-Central United States, the Californian Coast, and the South and East Coasts of the USA. In summer, decreases in the wind speed ranging from 5 to 10 percent per century are found over the same coastal regions. These projected changes in the surface wind speed are moderate overall. From the global model projections, the estimate of wind power potential for North America based on present-day climatology will remain accurate in the coming decades. The relatively coarse resolutions of the global models do not allow an accurate representation of the mesoscale and submesoscale topography, which might affect the projections of the changes in the surface wind field. Our results will serve as a useful basis to guide future work on downscaling the CMIP5 model outputs to the submesoscale, which may help resolve the topographic effects. The RCP scenarios used in CMIP5 do not consider the effects of future land-use changes, including those related to the construction of large-scale wind farms. An integration of those effects into regional climate modeling, using the CMIP5 global model outputs as the boundary conditions, will help refine the conclusions of this work.

REFERENCES

1. A. Lopez, B. D. Roberts, N. Heimiller, N. Blair, and G. Porro, "U.S Renewable Energy Technical Potential: A GIS based analysis," Tech. Rep. NREL/TP-6A20-51946, National Renewable Energy Laboratory, Golden, Colo, USA, 2012.
2. J. M. Freedman, K. T. Waight, and P. B. Duffy, "Does climate change threaten wind resources?" North American Wind Power, vol. 6, pp. 49–53, 2009.
3. D. Ren, "Effects of global warming on wind energy availability," Journal of Renewable and Sustainable Energy, vol. 2, no. 5, Article ID 052301, 2010.
4. K. E. Taylor, R. J. Stouffer, and G. A. Meehl, "An overview of CMIP5 and the experiment design," Bulletin of the American Meteorological Society, vol. 93, no. 4, pp. 485–498, 2012.
5. IPCC, Climate Change 2013: The Physical Science Basis, Working Group I Contribution to Fifth Assessment Report of Intergovernmental Panel on Climate Change, Cambridge University Press, Cambridge, UK, 2013.
6. R. Seager, M. Ting, I. Held et al., "Model projections of an imminent transition to a more arid climate in southwestern North America," Science, vol. 316, no. 5828, pp. 1181–1184, 2007.
7. N. C. Baker and H.-P. Huang, "A comparative study of precipitation and evaporation in semiarid regions between the CMIP3 and CMIP5 climate model ensembles," Journal of Climate, vol. 27, no. 10, pp. 3731–3749, 2014.
8. S. C. Pryor and R. J. Barthelmie, "Assessing climate change impacts on the near-term stability of the wind energy resource over the United States," Proceedings of the National Academy of Sciences of the United States of America, vol. 108, no. 20, pp. 8167–8171, 2011.
9. L. O. Mearns, R. Arritt, S. Biner et al., "The North American regional climate change assessment program: overview of phase I results," Bulletin of the American Meteorological Society, vol. 93, no. 9, pp. 1337–1362, 2012.
10. G. A. Meehl, C. Covey, T. Delworth et al., "The WCRP CMIP3 multimodel dataset: a new era in climatic change research," Bulletin of the American Meteorological Society, vol. 88, no. 9, pp. 1383–1394, 2007.
11. D. W. Keith, J. F. DeCarolis, D. C. Denkenberger et al., "The influence of large-scale wind power on global climate," Proceedings of the National Academy of Sciences of the United States of America, vol. 101, no. 46, pp. 16115–16120, 2004.
12. A. S. Adams and D. W. Keith, "Are global wind power resource estimates overstated?" Environmental Research Letters, vol. 8, Article ID 015021, 2013.
13. H. Paek and H.-P. Huang, "Centennial trend and decadal-to-interdecadal variability of atmospheric angular momentum in CMIP3 and CMIP5 simulations," Journal of Climate, vol. 26, no. 11, pp. 3846–3864, 2013.
14. M. Kaltschmitt, W. Streicher, and A. Wiese, Renewable Energy: Technology, Econom Ics, and Environment, Springer, New York, NY, USA, 2007.
15. M. Kanamitsu, W. Ebisuzaki, J. Woollen et al., "NCEP-DOE AMIP-II reanalysis (R-2)," Bulletin of the American Meteorological Society, vol. 83, no. 11, pp. 1631–1559, 2002.
16. E. W. Peterson and J. P. Hennessey Jr., "On the use of power laws for estimates of wind power potential," Journal of Applied Meteorology, vol. 17, no. 3, pp. 390–394, 1978.

CHAPTER 2

Potential Climatic Impacts and Reliability of Very Large-Scale Wind Farms

C. WANG AND R. G. PRINN

2.1 INTRODUCTION

World energy demand is predicted to increase from ~430 EJ/year (14 TW) in 2002 to ~1400 EJ/year (44 TW) in 2100 (Reilly and Paltsev, 2007). Any effective energy contributor needs to be implemented on a very large scale (e.g. provide 10% of the year 2100 demand). Among the current energy technologies with low or zero greenhouse gas (GHG) emissions, electrical generation using wind turbines is percentage-wise the fastest growing energy resource worldwide. In the US, it has grown from 1.8 GW of capacity in 1996 to more than 11.6 GW (~0.37 EJ/year) in 2006, but this is still negligible compared to future energy demand.

The solar energy absorbed by the Earth is converted into latent heat (by evaporation), gravitational potential energy (by atmospheric expansion), internal energy (by atmospheric and oceanic warming, condensation), or kinetic energy (e.g by convective and baroclinic instabilities) (Lorenz, 1967). Averaged globally, internal energy, gravitational potential energy, latent heat, and kinetic energy comprise about 70.4, 27.05, 2.5, and 0.05% respectively of the total atmospheric energy (Peixoto and Oort, 1992). However, only a small fraction of the already scarce kinetic energy is contained in the near surface winds that then produce small-scale turbulent motions due to surface friction. Eventually the turbulent motions downscale to molecular motions, thus converting bulk air kinetic energy to internal energy.

However, it is not the size of these energy reservoirs, but the rate of conversion from one to another, that is more relevant here. The global average rate of conversion of largescale wind kinetic energy to internal energy near the surface is about 1.68 W/m^2 (860 TW globally) in our model calculations. This is only about 0.7% of the average net incoming solar energy of 238 W/m^2 (122 PW globally) (Lorenz, 1967; Peixoto and Oort, 1992). The magnitude of this rate when wind turbines are present is expected to differ from this, but not by large factors. The widespread availability of wind power has fueled substantial interest in harnessing it for energy production (e.g. Carter, 1926; Hewson, 1975; Archer and Jacobson, 2003). Wind turbines convert wind power into electrical power. However, the turbulence near the surface, which also feeds on wind power, is critical for driving the heat and moisture exchanges between the surface and the atmosphere that play an important role in determining surface temperature, atmospheric circulation and the hydrological cycle.

Because of the low output (~MW) of individual wind turbines, one needs to install a large number of the devices to generate a substantial amount of energy. For example, presuming these turbines are effectively generating at full capacity only 1/3 of the time, about 13 million of them are needed to meet an energy output of 140 EJ/year (4.4 TW), and they would occupy a continental-scale area. While the amount of energy gained from global deployment of surface wind power may be small relative to the 860 TW available globally, the accompanying climate effects may not be negligible. A previous study using atmospheric general circulation models with fixed sea surface temperatures suggests that the climatic perturbation caused by a large-scale land installation of wind turbines can spread well beyond the installation regions (Keith et al., 2004).

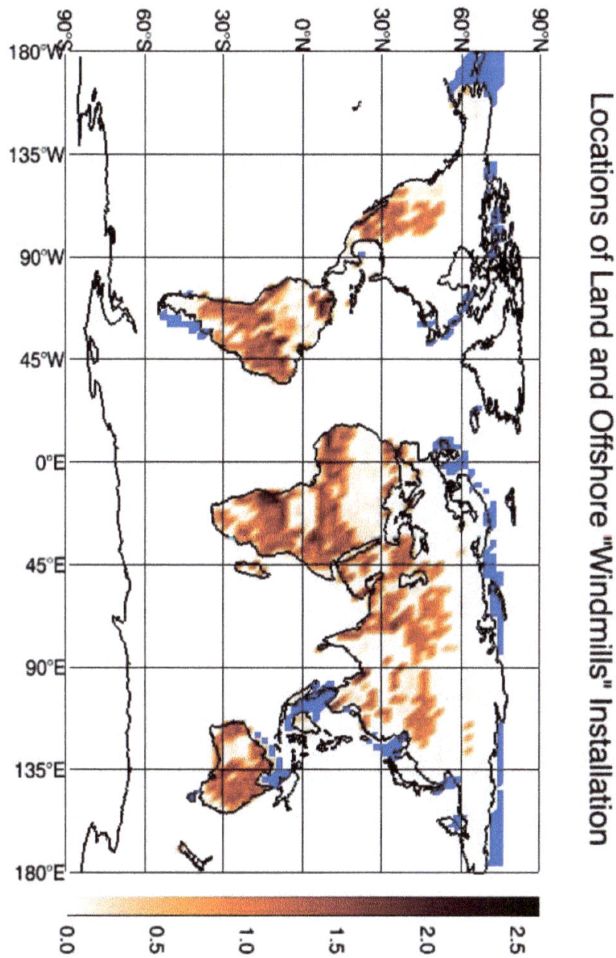

FIGURE 1: Locations of land installations are indicated by the modeled change of surface drag coefficient (non-dimensional) averaged over the final 20 years of the 60-year Run L (see color code on right hand side). The drag values have been scaled by a factor of 1000. Also shown are the locations of offshore installation regions where the ocean depth is shallower than 200 m (blue shading).

2.2 METHODS

To explore the potential climate impacts of very largescale windfarms, we use, for the first time, a fully coupled atmosphere-ocean-land system model, specifically the Community Climate Model Version 3 of the US National Center for Atmospheric Research with a mixed layer ocean (Kiehl et al., 1998). In order to isolate the climate effects of wind turbines from those due to greenhouse gas increases, all runs were carried out with current greenhouse gas levels. The chosen T42 spatial spectral resolution provides an approximately 2.8 by 2.8 degree grid point spacing in the horizontal, and there are 18 vertical layers.

Seven model runs with 60-year durations were carried out and are reported here. Each run takes about 40 years to reach climatic steady states that approximately repeat annually after that. Four of the five runs (denoted VL, L, H, and VH) used different schemes to simulate the wind turbine effects over land, while another run (REF) excludes any wind turbine effects and thus serves as the control or reference. Besides the land installation simulations, we have also conducted two additional runs (denoted OL and OH) in which we simulate installing wind turbines over all coastal regions between 60◦ S and 74◦ N in latitude where the ocean depth is shallower than 200 m. As before, comparisons of the oceanic wind turbine runs with the REF run serve to isolate the climate effects of the wind turbines. Unless otherwise indicated, the means of the last 20 years (years 41–60) of each of the model integrations are used in the analyses.

Previous model studies of wind farms of various scales have used methods to increase the surface roughness to simulate the aerodynamic effect of wind turbines (Frandsen, 1992; Baidya Roy et al., 2004; Keith et al., 2004; also see the review by Crespo et al., 1999). We adopt the same general approach, but use model-provided parameters for objects similar to wind turbines. We selected the global land regions covered by grass (including cold C3 and warm C4) and shrub (including evergreen and deciduous) to be the sites for installation of the wind turbines over the land. This choice is influenced by the generally lower economic value and high wind speeds over such lands, but future studies might investigate alternative strategies. The wind turbine effect is simulated specifically by modifying the model surface roughness and/or displacement height coefficients over the global grass and shrub regions in the land model of the CCM3 system. The selected roughness and displacement height in the four wind turbine runs are: Run VL, 0.12 m (double the original value) and 0.34 m (unchanged); Run L, 0.16 m (arbitrary) and 0.34 m; Run H, 0.75 m (arbitrary, close to the value of 0.77 m of the needle leaf deciduous tree in the model) and 0.34 m; and Run VH, 2.62 m and 23.45 m (based on the evergreen forest in the model). In the ocean-based experiments, an additional surface drag of 0.007 and 0.001 over the installed regions has been applied in the Runs OH and OL, respectively, to simulate the wind turbine effect on wind power extraction. The former value is about the same as a reported measurement over mesoscale windfarms (see Keith et al., 2004) while the latter is about double the average sea surface roughness (Peixoto and Oort, 1992). Note that the equations describing the atmosphere-ocean interfacial interactions in the model are highly parameterized and defining a formulation to mimic wind turbines with equivalent realism to the one used for the land-based experiments is difficult. Therefore, the two ocean experiments are for exploratory purposes only.

Except for the changes made to the surface roughness or displacement height described above, we keep all other surface properties in these regions identical to their standard CCM3 settings. The model calculates the actual surface properties based on weighted values over all surface types in a given grid. Our method for land installations avoids changing uniformly the above two surface properties of a given model grid to those of a modeled wind farm unless one or both of the two selected surface types (grass, shrub) dominate the grid.

The rate of conversion of large-scale kinetic energy to turbulent kinetic energy can be described by a term in the equation for the change in the mean flow kinetic energy per unit volume of air (KE) with time t (Stull, 1988):

$$\frac{dKE}{dt} = -\rho u_i' u_j' \frac{\partial U_i}{\partial x_j} = \tau_{i,j} \frac{\partial U_i}{\partial x_j} \qquad (1)$$

Here, i,j = 1, 2, 3 are the three directions of the spatial coordinates, x, U is the mean wind speed, u' is the deviation of the actual wind from the mean (so that it reflects the turbulent motions), ρ is the air density, and τ is the surface stress. The same term exists in the equation of change of turbulent kinetic energy but with an opposite sign. The surface stress is derived in the land surface model (LSM) or the mixed-layer ocean model of CCM3 as a function of surface properties including roughness and displacement height. The change in the rate of downward transport of cascaded kinetic energy due to the simulated wind turbine effects are calculated continuously at each model time step (20 min) by comparing the surface stress values derived with and without the perturbed surface roughness and/or displacement, respectively. These calculated changes are then used to calculate the uptake of wind power by the simulated wind turbines which is then partially converted to the actual electrical power output.

The various changes in surface properties lead to an increase of surface momentum drag and a decrease of local near-surface wind speed. The changes in surface momentum drag in Run L were up to 0.0025, depending on the dominance of the grass and shrub types in the given model grid (Fig. 1). This effect is enhanced in Runs H and VH, and reduced in VL. We install the wind turbines over 58 million km^2 of shrub and grass lands in the major continents, or over 10 million km^2 of the global coastal oceans where the depths are less than 200 m (Fig. 1). The relevant model parameters were derived in short trial runs to remove the amount of near-surface atmospheric kinetic energy needed to match various energy production targets. We make no specific assumptions about the type and spacing of these wind turbines; our interest is only in determining the impacts of removing the kinetic energy from the near-surface atmosphere needed to drive them.

2.3 RESULTS

In Run L with a moderate change in the surface roughness over the installed land regions, the reduction of wind power due to the wind turbines is about 20 TW, or 630 EJ/yr, which is about 2.3% of the total rate of conversion of mean flow to turbulent kinetic energy at the Earth's surface and 23% of the conversion rate over the actual areas of the wind turbine installation. No more than 59% of the kinetic energy contained in an airstream tube having the same cross section as a disc-shaped obstacle can be converted to useful work by the disc (the Lanchester-Betz-Joukowsky limit) (van Kulk, 2007). The actual conversion efficiency of this kinetic energy to electric power is likely to be lower than 30% (Busawon et al., 2005). With a conversion efficiency of 25%, the wind turbines in Run L would provide about 158 EJ/yr (5 TW). In three other numerical experiments, the kinetic energy extracted by the wind turbines was either reduced or enhanced compared to Run L (e.g. reflecting the effects of lowering or raising the wind turbine spatial density). The computed electrical energy outputs are about 72, 344, and 603 EJ/yr (2.3, 11 and 19 TW) in Run VL, H, and VH, respectively. The offshore shallow ocean installations provide about 96 and 30 EJ/yr (3.0 and 0.95 TW) in Run OH and OL, respectively.

The computed air temperature over the installation regions in Run L is elevated by more than 1 ◦C in the lowest model layer (~30 m thick at sea level) in many regions (Fig. 2), but the increase, averaged over the entire global land surface, is only about 0.15 ◦C. Although the surface air temperature change is dominated by the increase over the wind turbinein-stalled areas (Fig. 1), the changes go well beyond these areas (Fig. 2). The frequency distributions for temperature changes for Run L and the other three land-based runs, are also shown in Fig. 2. The global land-average temperature changes are 0.05, 0.16, and 0.73 ◦C, respectively, for these three other land-based runs (VL, H and VH). In all these runs, except for Run VL, the global patterns of these changes are consistent with Run L (Fig. 2). These patterns also have some similarities to the previous study by Keith et al. (2004) over land, but not over the oceans, since that study assumed fixed ocean temperatures.

The warming caused by the wind turbines is limited to the lowermost atmospheric layers (Fig. 3). Above the planetary boundary layer, a compensating cooling effect is expected and observed in many regions, because the turbulent transfer of heat from the surface to these higher layers is reduced. This should be contrasted to the relatively uniformly distributed warming throughout the troposphere induced by rising greenhouse gases (IPCC, 2007).

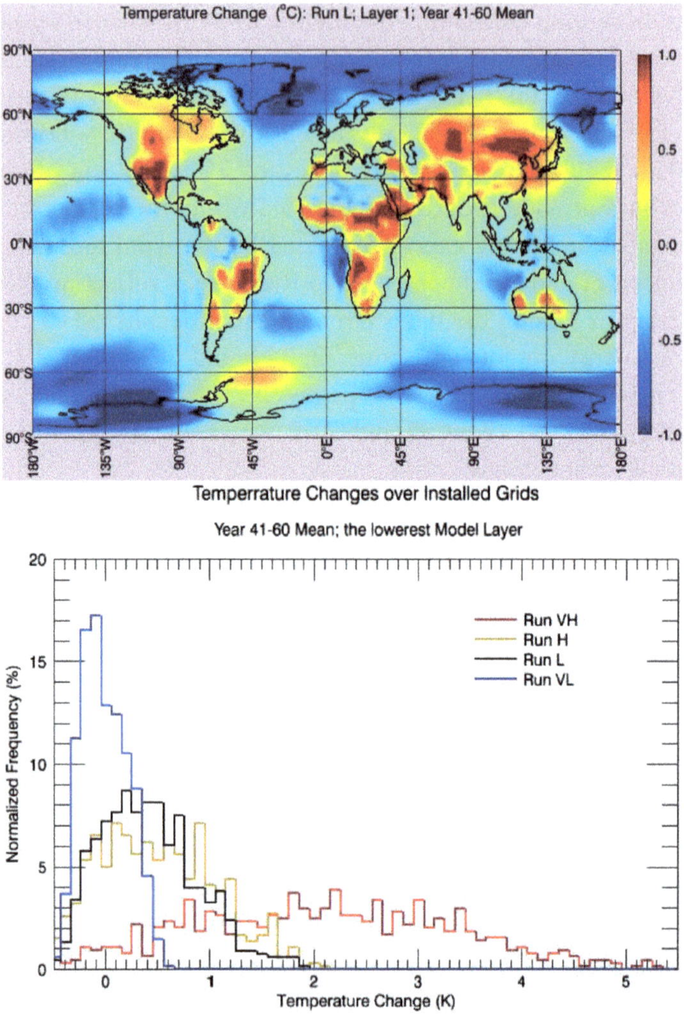

FIGURE 2: Temperature changes (Run L minus reference, REF) in the lowest model layer resulting from large-scale deployment of wind turbines over land sufficient to generate 158 EJ/year of electric power (upper panel); and normalized frequency of temperature changes over the installation regions in Runs VH, H, L, and VL (lower panel). Both refer to averages over years 41–60.

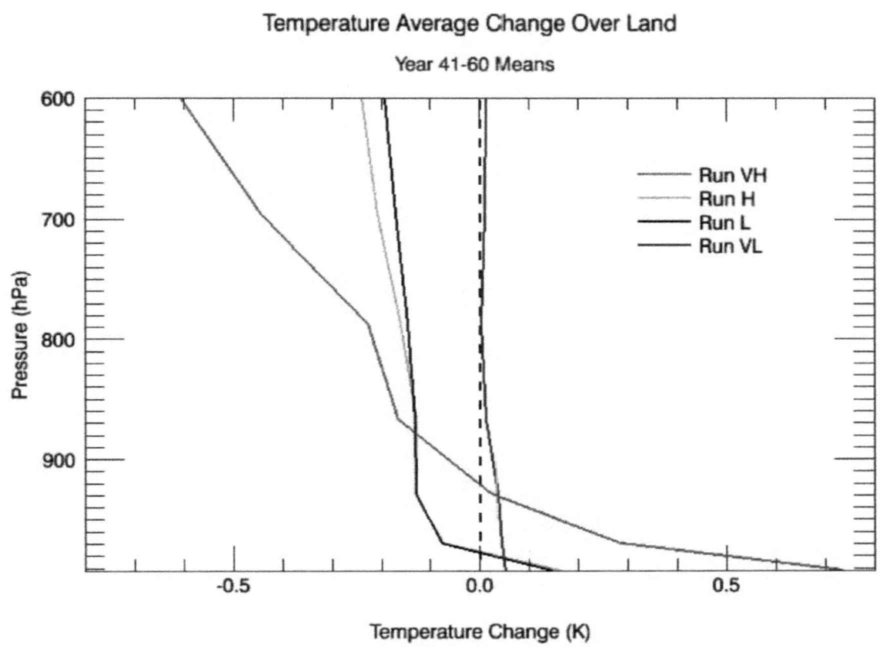

FIGURE 3: Horizontally averaged temperature changes (relative to the reference, REF) over land in the 4 wind turbine installation runs. All data are 20-year means from year 41 to 60.

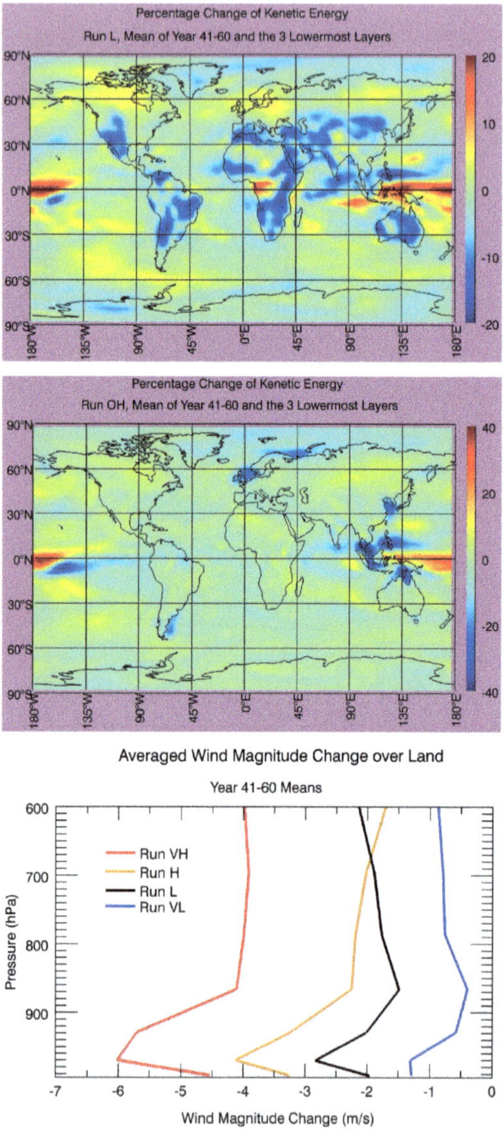

FIGURE 4: The changes in the kinetic energy per unit mass are shown in the upper panel for Run L, and the middle panel for Run OH. Also shown in the lower panel are the horizontally averaged wind magnitude changes (derived from the kinetic energy per unit mass change) over land in the 4 wind turbine installation runs. All data are 20-year means from year 41 to 60.

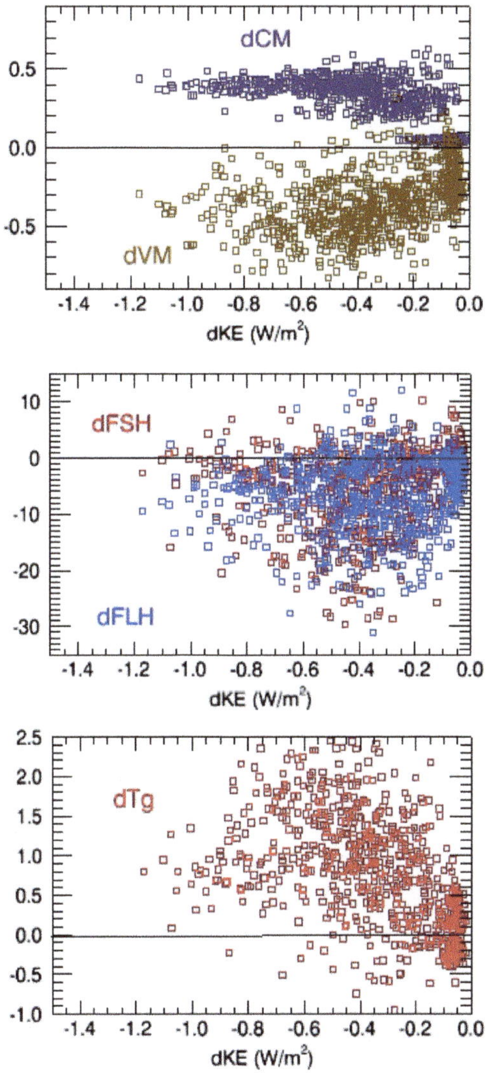

FIGURE 5: Changes in surface momentum drag coefficient (dCM; unitless), wind magnitude (dVM; m/s), sensible (dFSH; W/m^2) and latent (dFLH; W/m^2) heat fluxes, and surface air temperature (dTg; K) over the model grids where the kinetic energy losses (dKE; W/m^2) due to wind turbines occur. Results shown are year 41–60 means of Run L minus Run REF.

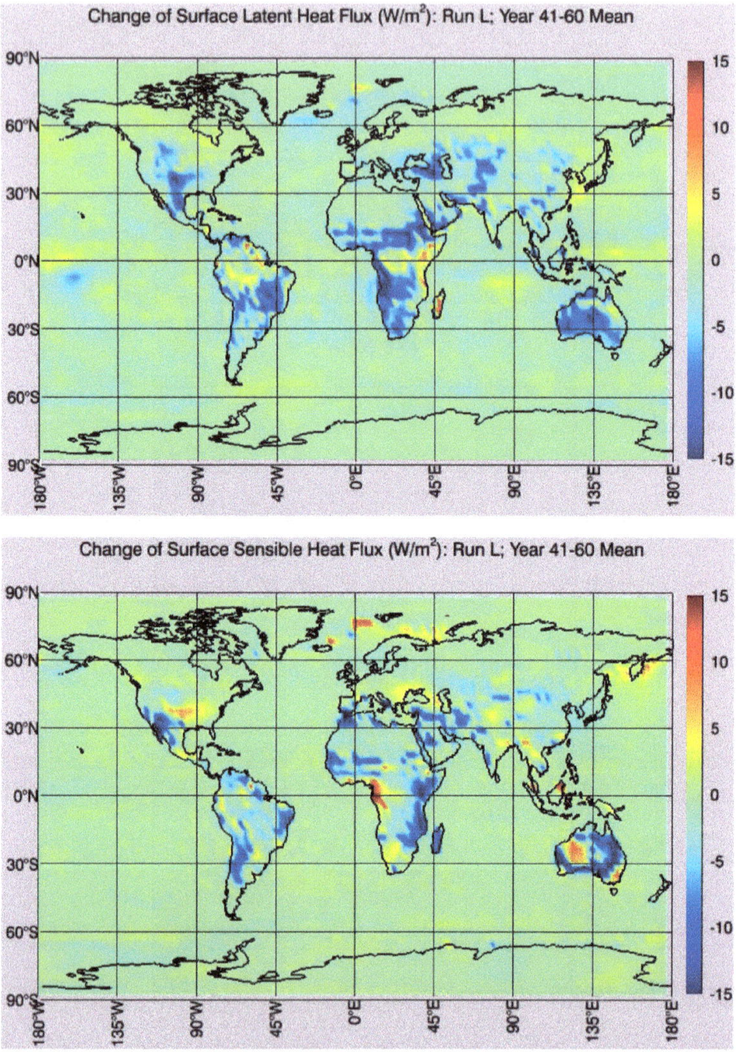

FIGURE 6: Surface heat flux changes for Run L (Run L minus Run REF) for latent heat (upper panel) and sensible heat (lower panel) fluxes. Both panels are in W/m^2 and averaged over years 41–60.

Increasing surface roughness (to simulate the wind turbines) without significantly lowering the near-surface wind speed should increase near-surface turbulent latent and sensible heat transport and thus cool the surface. However, changes in surface roughness over a region with a very large width in the prevailing wind direction indeed cause a significant reduction in the wind speed. Our results suggest that the latter effect prevails over the majority of the installation regions (Fig. 4). When averaged over the land, the reductions of wind magnitude in the various land-installation cases display a similar vertical profile, with a peak in the second model layer above the surface (Fig. 4c). Note that the difference between Run L and Run H is defined more clearly in the wind reduction than in the temperature change (Fig. 3), implying that the temperature change is more closely related to the vertical turbulent heat exchange. The effect of wind reduction over the installed land regions appears to exceed the effect of the surface roughness increase. The wind reduction specifically weakens the near-surface vertical turbulent transport, and thus warms the surface (Fig. 5). Note that, like the effects on temperature, the effects of these wind turbine installations on wind magnitude, surface heat fluxes, spread well beyond the installation regions and often have opposite signs to those in the installation regions (Figs. 4 and 6). These long-range effects are likely to be very model-dependent. Dynamical mechanisms involving Rossby waves for long-range effects of large-scale changes in land surface friction have been proposed (Kirk-Davidoff and Keith, 2008). Long-range effects are also computed in climate model simulations where regional energy budgets are altered by aerosols (Wang, 2007).

Note that the fractional changes of surface drag in our two ocean-based runs using a slab ocean model are very high compared to the land cases, owing to the much higher intrinsic surface roughness over land than over ocean (Fig. 7). Therefore, in contrast to the land-based experiments, this substantial increase in surface drag in the ocean-based experiments creates much stronger turbulence that substantially opposes the wind reduction effect due to the roughness change. This leads to an enhancement in ocean-atmosphere heat fluxes, particularly latent heat fluxes, and thus to local cooling over almost all of the installation regions (Fig. 8). As in the land-based runs, the temperature changes in the coastal ocean-based runs also occur well beyond the installation regions with similar vertical profiles (not shown) to Fig. 3, but with opposite signs. Note that these results in these two ocean runs are likely not reliable, since they are dependent on the single CCM3 model option available to us for making the changes to ocean surface properties necessary to simulate the drag effects of wind turbines over water. These results might also differ from those derived using a full ocean GCM that allows changes in oceanic circulation.

The spatially extensive changes in temperatures and surface heat fluxes for the land installations are sufficient to affect the global distributions of cloud cover, especially the lower clouds (not shown), and precipitation (Fig. 9). The rates of convective precipitation (Fig. 9) are generally reduced in the Northern Hemisphere and enhanced in the Southern Hemisphere, symptomatic of a shift in the atmospheric Hadley Circulation (Wang, 2007, 2009), reflecting an alteration to the large-scale circulation by the surface roughness changes caused by wind turbines (see e.g. Kirk-Davidoff and Keith, 2008). In the mid-latitudes, especially in the Northern Hemisphere, changes in large-scale precipitation also appear (Fig. 9), indicating an impact on mid-latitude weather systems. Although the changes in local convective and large-scale precipitation exceed 10% in some areas, the global average changes are not very large.

To investigate the issue of wind variability leading to intermittency in wind power generation, we show in Fig. 10 the average and standard deviation of the monthly-mean wind power consumption (DKE=dKE/dt in TW, see Eq. 1) for each month of the year and for each continent over the last 20 years of Run L. Also shown is the time series of these monthly-means over the last 20 years. Dividing DKE by 4 for a 25% conversion efficiency, the 20-year average generated electrical power over each continent is 0.57 (North America), 0.72 TW (South America), 1.28 TW (Africa/Middle East), 0.63 TW (Australia), and 1.29 TW (Eurasia). However, quite apart from the well-known day-to-night and day-to-day intermittency of wind turbines, from Fig. 10 there are very large (up to a factor of 2) and geographically extensive seasonal variations especially over North and South America and Africa/Middle East. Unfortunately the months of minimum generation usually coincide with maximum demand for air conditioning. In an electrical generation system dominated by wind turbines, reliability of supply cannot therefore be achieved simply by long-distance power transmission over these continents.

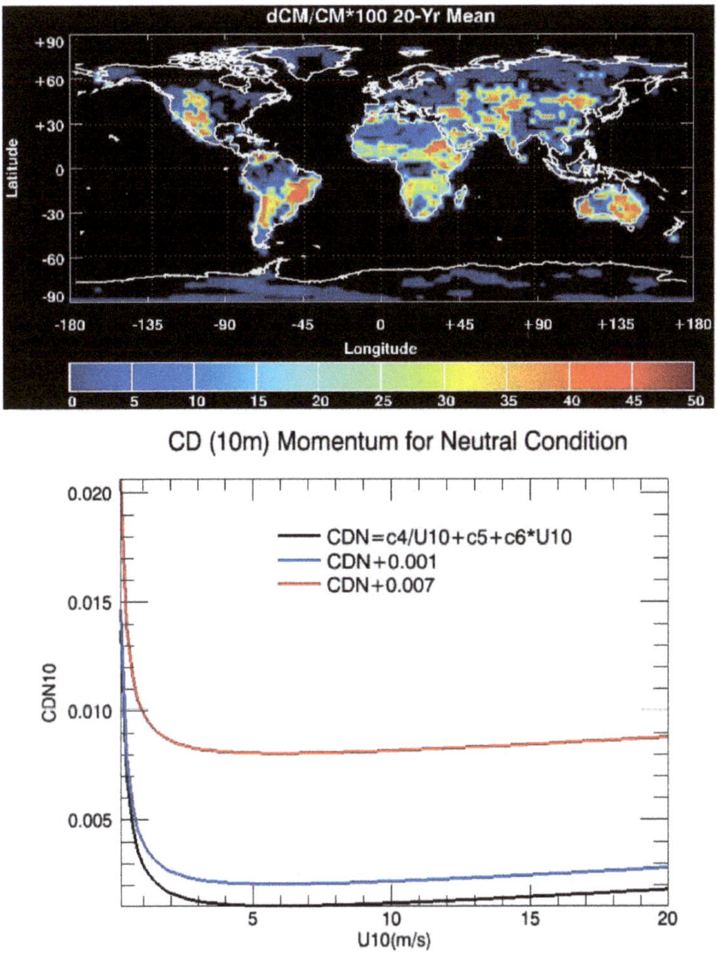

FIGURE 7: Percentage changes of surface momentum drag coefficient (dCM/CM*100) due to the simulated wind turbines over the land (upper panel); and surface momentum coefficients (CDN10) without (black line) and with (Run OL, blue line; Run OH, red line) the simulated wind turbines over the ocean (lower panel).

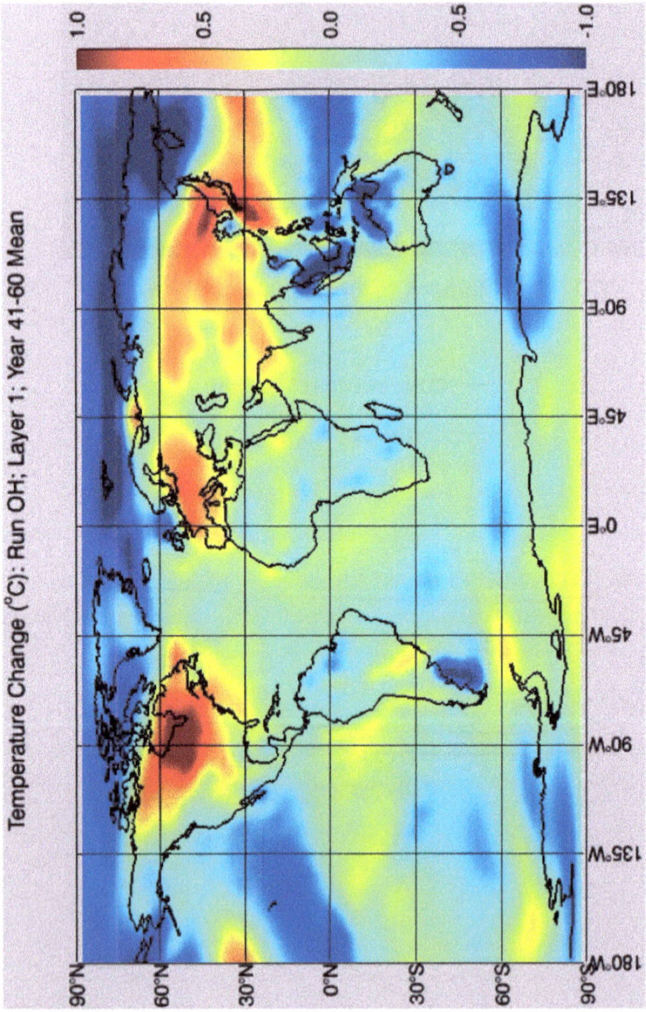

FIGURE 8: Same as Fig. 2 (upper panel) but for Run OH.

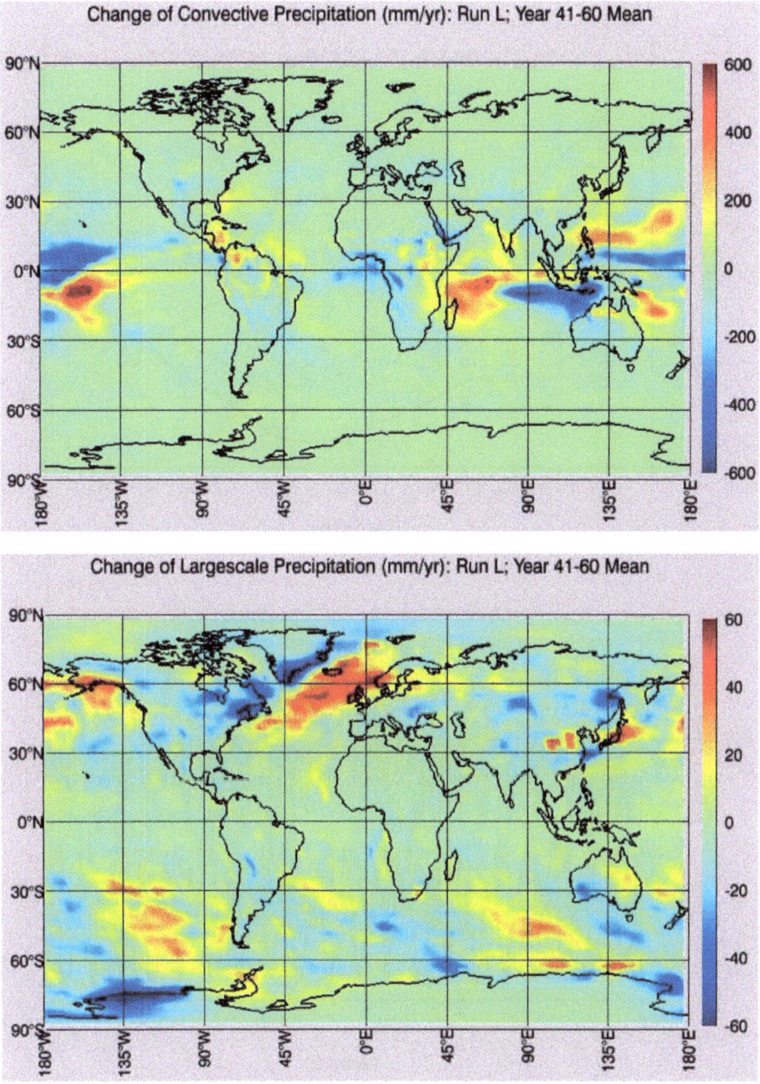

FIGURE 9: Precipitation changes (Run L minus Run REF) for: convective precipitation (upper panel), and large-scale precipitation (lower panel). Both are in mm/yr and averaged over years 41–60.

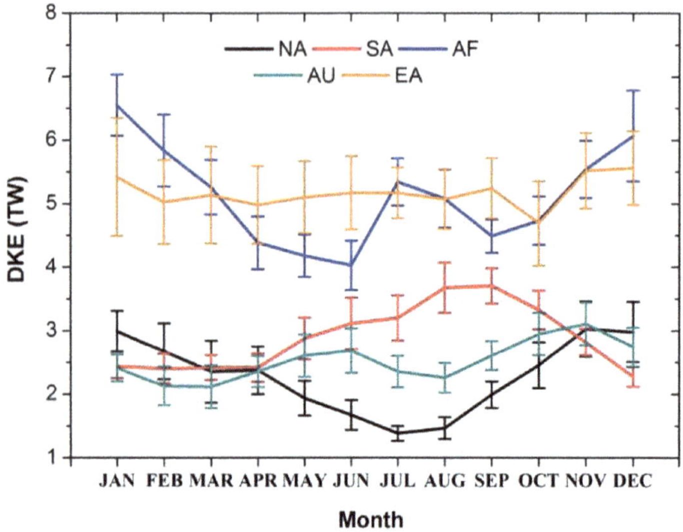

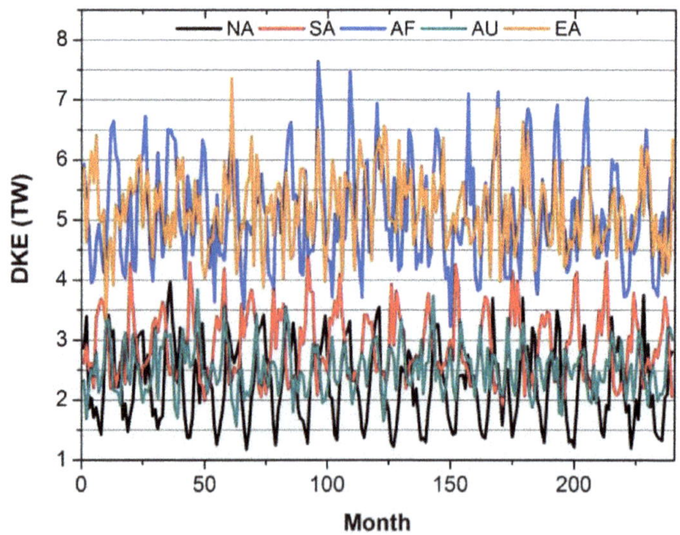

FIGURE 10: Twenty-year (years 41–60) averages and standard deviations (upper panel), and all values (lower panel), of the monthly mean wind power consumption (DKE=dKE/dt, Eq. 1) by simulated wind turbines installed in various continents: North America (NA), South America (SA), Africa and Middle East (AF), Australia (AU), and Eurasia (EA).

2.4 CONCLUSIONS

Meeting future world energy needs while addressing climate change requires large-scale deployment of low or zero GHG emission technologies such as wind energy. We used a threedimensional climate model to simulate the potential climate effects associated with installation of wind-powered generators over vast areas of land and ocean. Using wind turbines to meet 10% or more of global energy demand in 2100 could cause surface warming exceeding 1 °C over land installations. Significant warming and cooling remote from the installations, and alterations of the global distributions of rainfall and clouds also occur.

Our ocean results indicating cooling over the installation regions and warming and cooling elsewhere are interesting, but suspect due to the unrealistic increases in surface drag needed to extract the target wind power. Specific new and realistic parameterizations for simulating the effects of wind turbines over the ocean will need to be developed and applied in general circulation models before reliable conclusions can be reached. Future advances in floating turbine technology might enable the installation of wind turbines over oceans at depths exceeding our assumed 200-m maximum depth range, presuming that the issue of the needed long-range electricity transmission is resolved.

Installation of wind turbines over land areas that have alternative spatial extents, topographies and hydrological properties would produce different, but presumably still significant, climate effects. Due to the computed nonlinearity between the changes in surface roughness and the climate response, defining the optimal deployment of wind turbines is challenging. Climatic effects increase with power generated and decrease with conversion efficiency, putting aside the potential environmental effects for instance on birds and weather radar as well as on ambient noise levels. Also, for the widely spaced wind turbines simulated in our runs, the environmental effects appear small when they are generating less than 1 TW globally even with current technologies.

Our results should be fairly robust to assumptions about the specific wind turbine technologies utilized. Increasing their efficiencies from 25% to 35% helps to lower, but does not remove the calculated climate effects. Our results are dependent upon the realism of the land surface and atmospheric boundary layer in our chosen climate model, and investigations with alternative models, including higher-resolution climate models with fully dynamical three-dimensional oceans are warranted. Our method involving varying the surface roughness has been shown to capture certain features of the effects of turbines on the local wind using mesoscale models (Frandsen, 1992; Vermeer et al., 2003). However, this method cannot explicitly resolve the detailed vertical wind profiles affected by atmospheric stability or wind shear that are clearly subgrid scale processes in our model (Vermeer et al., 2003; Lange and Focken, 2005). Appropriate field experiments to test our conclusions, and to explore better ways for simulating wind turbines in models, are also required.

Finally, intermittency of wind power on daily, monthly and longer time scales as computed in these simulations and inferred from meteorological observations, poses a demand for one or more options to ensure reliability, including backup generation capacity, very long distance power transmission lines, and on-site energy storage, each with specific economic and/or technological challenges

REFERENCES

1. Archer, C. L. and Jacobson, M. Z.: Spatial and temporal distributions of US winds and wind power at 80 m derived from measurements, J. Geophys. Res., 108, 4289, doi:10.1029/2002JD002076, 2003.
2. Baidya Roy, S., Pacala, S. W., and Walko, R. L.: Can large wind farms affect local meteorology?, J. Geophys. Res., 109, D19101, doi:10.1029/2004JD004763, 2004.
3. Busawon, K., Dodson, L., and Jovanovic, M.: Estimation of the power coefficient in a wind conversion system, Proc. 44th IEEE Conf. on Decision and Control, and the European Control Conf. 2005, Seville, Spain, 12–15 December, 3450–3455, 2005.
4. Carter, H. G.: Wind as motive power for electrical generators, Mon. Wea. Rev., 54, 374–376, 1926.
5. Crespo, A., Hernandez, J. H., and Frandsen, S.: Survey of modeling methods for wind turbine wakes and wind farms, Wind Energy, 2, 1–24, 1999.
6. Frandsen, S.: On the wind speed reduction in the center of large clusters of wind turbines, J. Wind Eng. Indus. Aerodyn., 39, 251– 265, 1992.
7. Hewson, E. W.: Generation of power from the wind, Bull. Amer. Meteor. Soc., 56, 660–675, 1975.
8. IPCC: Climate Change 2007: The Physical Science Basis, Contribution of Working Group I to the Fourth Assessment Report of the Intergovernmental Panel on

Climate Change, edited by: Solomon, S., Qin, D., Manning, M., Chen, Z. Marquis, M., Averyt, K. B., Tignor, M., and Miller, H. L., in: Cambridge University Press, Cambridge, UK and New York, NY, USA, 996 pp., 2007.

9. Keith, D. W., DeCarolis, J. F., Denkenberger, D. C., Lenschow, D. H., Malyshev, S. L., Pacala, S. N., and Rasch, P. J.: The influence of large-scale wind power on global climate, P. Natl. Acad. Sci. USA, 101, 16115–16120, 2004.
10. Kiehl, J. T., Hack, J. J., Bonan, G. B., Boville, B. A., Williams, D. L., and Rasch, P. J.: The National Center for Atmospheric Research Community Climate Model: CCM3, J. Climate, 11, 1131– 1149, 1998.
11. Kirk-Davidoff, D. B. and Keith, D. W.: On the Climate Impact of Surface Roughness Anomalies, J. Atmos. Sci., 65, 2215–2234, doi:10.1175/2007JAS2509.1, 2008.

PART II

SOLAR ENERGY AND CLIMATE CHANGE

CHAPTER 3

Trends in Downward Solar Radiation at the Surface over North America from Climate Model Projections and Implications for Solar Energy

GERARDO ANDRES SAENZ AND HUEI-PING HUANG

3.1 INTRODUCTION

The last decade has witnessed a rapid development in solar energy as an alternative to fossil-fuel based energy. Solar power plants with increasing size and efficiency have been built. With an increased stake in the investment and return, site selection and assessments of long-term sustainability for solar power plants become increasingly important. One of the factors that affect the long-term planning for solar energy is the local climatology. Long hours of sunshine at a location are essential for a viable solar power plant. The available solar energy at a given site is quantified by the downward solar (shortwave) radiation at the surface. At a given latitude and day of the year, this quantity is affected by atmospheric water vapor and trace gases, the amount of aerosols in the atmosphere, and, most importantly, cloud cover (e.g., Li et al. [1]). Considering those factors, climatological maps of downward solar radiation have been widely produced for solar energy applications (e.g., National Renewable Energy Laboratory, http://www.nrel.gov/gis/solar.html, Maxwell et al. [2], and George and Maxwell [3]). Since climate is constantly changing due to anthropogenic and natural processes, the estimates of solar power potential based on present-day climatology are not guaranteed to be true in the future. In this study, we will analyze the projection of the changes in the downward solar radiation in the 21st century over North America using a set of climate model simulations driven by anthropogenic greenhouse-gas (GHG) forcing from the Climate Model Intercomparison Project—Phase 3 (CMIP3) archive (Meehl et al. [4]). The global climate models have relatively coarse horizontal resolutions but are capable of producing the first-order features of atmospheric general circulation. Over North America, GHG-induced changes in the large-scale circulation are known to produce future drying in the Southwest USA and a poleward shift of storm tracks over Western USA (e.g., Seager et al. [5] and Baker and Huang [6]). These changes potentially imply more sunshine in the Southwest USA but reduced sunshine in the higher latitudes in Western USA due to increased cloudiness associated with storms. We will quantify the extent to which these changes in atmospheric processes affect the downward solar radiation at the surface, as directly calculated by the climate models using their physical parameterization schemes.

For solar energy applications, it is relevant to know not only the changes in the seasonal mean solar radiation but also how these changes are distributed through different times of the day. The analysis of the latter requires subdaily data of solar radiation, which were archived only by a small number of modeling groups in CMIP. Nevertheless, with the limited data, we will make a first attempt to quantify the trends at different times of the day.

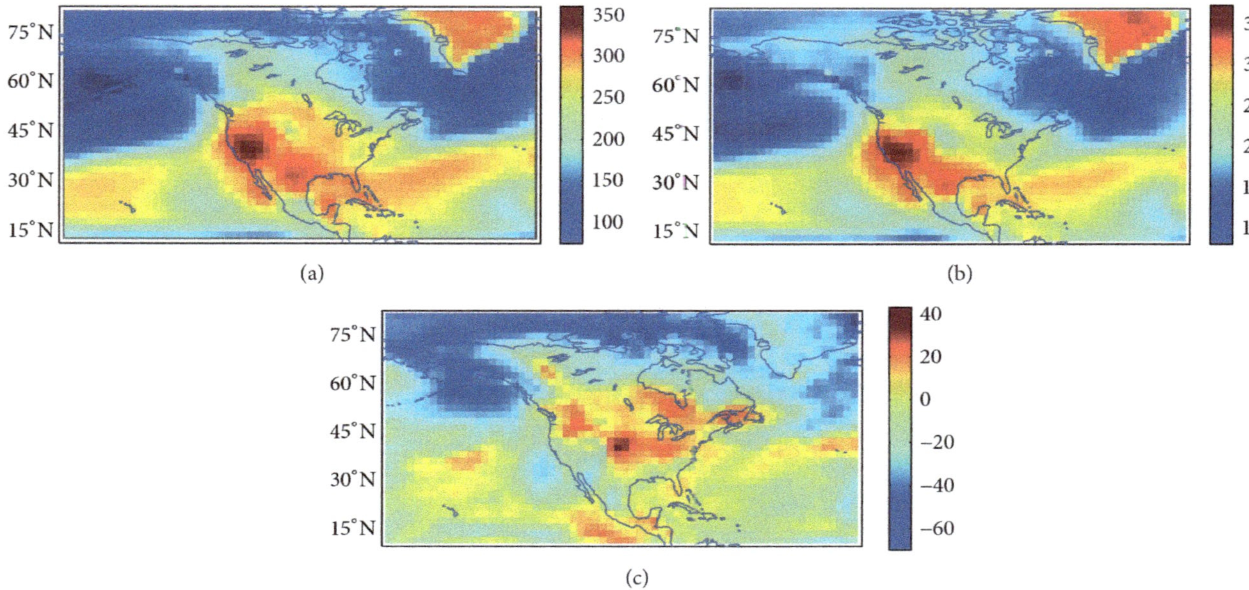

FIGURE 1: (a) The July climatology of downward shortwave radiation at the surface over North America from the 21st century run with GFDL CM2.0, based on 2080–2100 average. (b) The same as (a) but from the 20th century run, based on 1980–2000 average. (c) The trend, defined as (a) minus (b). The color scales with units in W/m^2 are shown on the right.

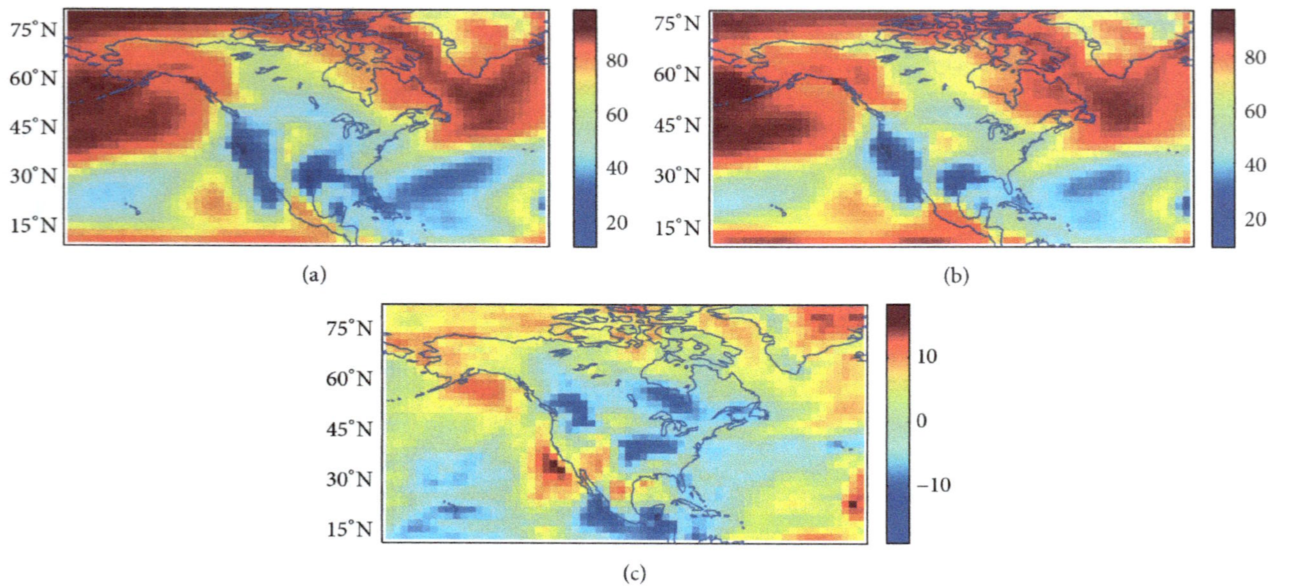

FIGURE 2: (a) The July climatology of total cloud fraction from the 21st century run with GFDL CM2.0, based on 2080–2100 average. (b) The same as (a) but from the 20th century run, based on 1980–2000 average. (c) The trend, defined as (a) minus (b). The color scales with units in percentage are shown on the right.

3.2 DATA FROM CLIMATE MODEL SIMULATIONS

We will analyze two sets of climate model simulations from the CMIP3 archive (http://www-pcmdi.llnl.gov/ipcc/about_ipcc.php). Although a large collection of model outputs have been archived by CMIP3 and the more recent CMIP5 (Taylor et al. [7]), the majority of the data are monthly means. Subdaily outputs are archived by only a small number of modeling groups and for short time periods. In this study, we choose to analyze the GFDL CM2.0 and MRI CGCM2.3.2 simulations in CMIP3. Both groups provided subdaily (3-hourly) archives of downward shortwave radiation at the surface for selected time slices in late 20th century and late 21st century. The surface radiation budget for CMIP3 20th century simulations was analyzed and compared to observation by Wild [8]. Over the midlatitude belt from 30 N to 60 N, the MRI model has about +15 W/m2 bias while the GFDL model has about −5 W/m^2 bias in the all-sky downward shortwave radiation at the surface (see Figure 6 in Wild [8]).

For both models, the GHG-induced trend will be deduced from the difference between the SRES A1B run (with increasing GHG concentration according to the A1B scenario) for the 21st century and the 20C3M run for the 20th century. We will first use the monthly mean archives to calculate the centennial trend, defined as the climatology of 2080–2100 minus the climatology of 1980–2000. For the analysis of the trends at different times of the day, we will use the 3-hourly model outputs as available from CMIP3. Each modeling group only provided the high-frequency data for short time slices in the late 20th century and late 21st century. Based on the availability of data, we will use the difference between year 2100 (from SRES A1B runs) and year 2000 (from 20C3M runs) to calculate the trends at selected local times during the day in North America.

This study will focus on the climate trend over North America. Over the Eurasian continent, Meleshko et al. [9] have shown that CMIP3 models project a 2–4% increase in winter and 2–10% decrease in summer of cloud cover (total cloud fraction) over Russia. The latter corresponds to an increase of 4-5 W/m^2 in summer in the area-averaged downward shortwave radiation at the surface. Since the populous regions in North America are located at lower latitudes than Russia, a similar change in cloud cover over

North America is expected to produce a greater change in the downward shortwave radiation at the surface.

3.3 RESULTS AND DISCUSSION

3.3.1 MONTHLY MEAN CLIMATOLOGY AND TREND

Figures 1(a) and 1(b) show the July climatology of the downward shortwave radiation at the surface over North America from GFDL CM2.0 simulations. Figure 1(a) is the 2080–2100 average from the SRES A1B run and Figure 1(b) is the 1980–2000 average from the 20C3M run. Figure 1(c) shows the trend deduced from the difference (future minus present) between Figures 1(a) and 1(b). For the climatology in Figures 1(a) and 1(b), although the downward solar radiation is zonally uniform at the top of the atmosphere, it becomes significantly nonuniform upon reaching the surface. This longitudinal nonuniformity is strongly influenced by cloudiness, as can be readily seen in Figure 2, the counterpart of Figure 1 for the total cloud fraction in July from the same simulations. For example, the strong downward shortwave radiation at the surface over Western and Southwest USA corresponds to the mostly clear-sky condition in July in those regions. In Figure 1(c), the model projected an overall positive trend over most of the United States, except for a small area in the Southwest USA. The increased sunlight at the surface, on the order of about 20 W/m2 (over the 21st century) or close to 10% of the climatology, is related to a reduced cloudiness over the USA (Figure 2(c)).

Figure 3 is similar to Figure 1 but for January. The trend in winter (Figure 3(c)) is an increase of the downward shortwave radiation over Southern and Southwest USA and a decrease of it over the northern half of the USA. The increased sunlight over Southwest USA in the cold season is consistent with the known projection by climate models of a drying trend in that region (Seager et al. [5] and Baker and Huang [6]). This drying trend is related, in part, to the poleward shift of storm tracks (Seager et al. [5]), which is consistent with the decrease in sunlight over Northern USA since the increase in storm activities implies an increase in cloudiness.

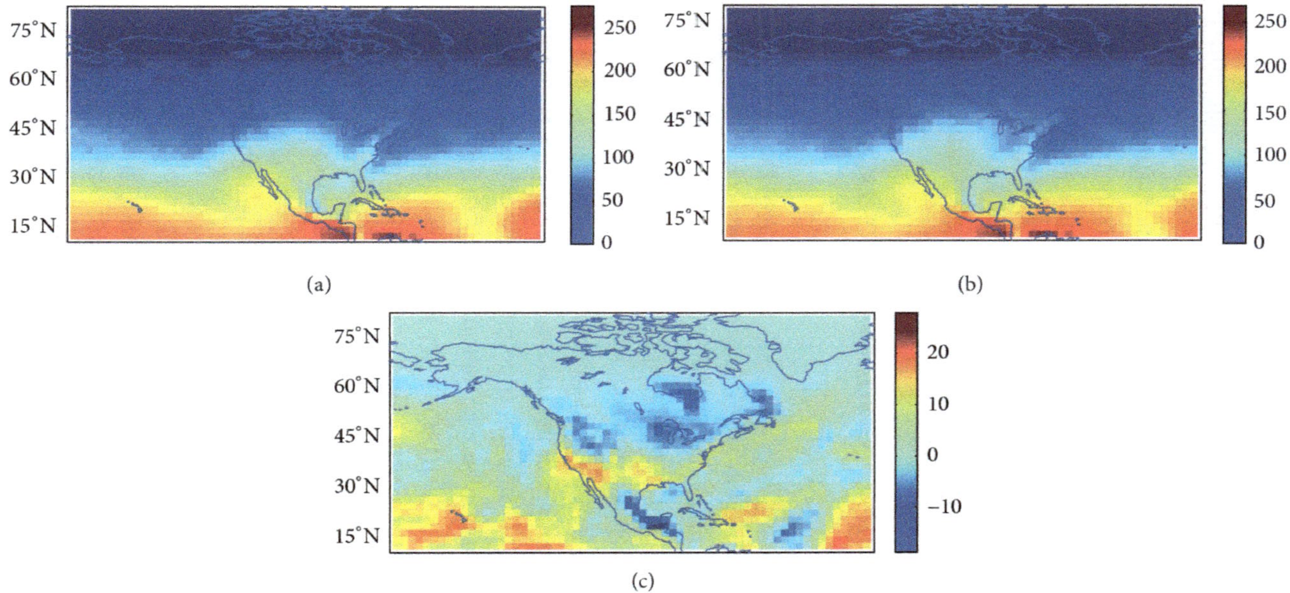

FIGURE 3: (a) The January climatology of downward shortwave radiation at the surface from the 21st century run with GFDL CM2.0, based on 2080–2100 average. (b) The same as (a) but from the 20th century run, based on 1980–2000 average. (c) The trend, defined as (a) minus (b). The color scales with units in W/m^2 are shown on the right.

The downward shortwave radiation at the surface simulated by MRI CGCM2.3.2 is shown in Figure 4. For conciseness, only the trends are shown. Figures 4(a) and 4(b), for July and January, respectively, are the counterparts of trends in Figures 1(c) and 3(c). In summer, the trend simulated by the MRI model is significantly different from that produced by the GFDL model. While both models project an increase of sunlight over Eastern USA and the Pacific Northwest, MRI projects a neutral to slightly negative trend over Western USA compared to a positive trend by the GFDL model. The trend in winter is more robust. Both models produced a decrease of sunlight over Northern USA and an increase of it over Southern USA. For both models, the magnitude of these trends is on the order of 10 W/m^2, or about 10% of the January climatology. As a notable difference, over the Southwest USA, GFDL CM2.0 produced a positive trend while MRI CGCM2.3.2 produced a neutral to negative trend. Since the trend in the shortwave radiation is strongly influenced by the trend in cloudiness which is highly parameterized in climate models, the differing projections by the two models are not surprising. A better agreement in the trend is found in January, possibly because the decrease in sunlight over Northern USA in the cold season is related to the poleward shift of storm tracks under an increasing GHG forcing, a phenomenon that is large scale in nature and is robustly simulated by the majority of climate models in CMIP3 (e.g., Yin [10]).

3.3.2 TRENDS AT DIFFERENT TIMES OF THE DAY

We next use the more limited data of 3-hourly model outputs to deduce the trends in the downward shortwave radiation at the surface as a function of the times of the day. Both GFDL CM2.0 and MRI CGCM2.3.2 provide the 3-hourly archives for the downward shortwave radiation for the year 2000 (from the 20C3M runs) and 2100 (from the SRES A1B runs). Figure 5 illustrates a 3-hourly time series of the downward shortwave radiation at the surface averaged over multiple grid points in Northern Arizona, for the year 2000, from the 20C3M simulation with GFDL CM2.0. The trend (now defined as the 2100 average minus the 2000 average) discussed in this section will be for a specific time of the day, averaged over either July or January.

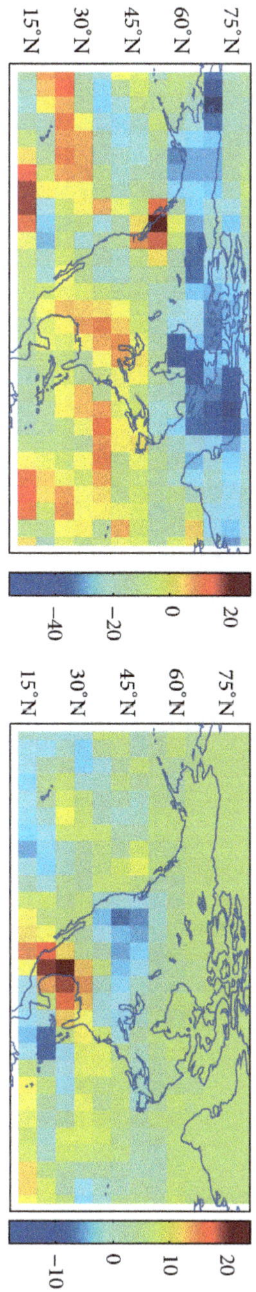

FIGURE 4: (a) The trend of the downward shortwave radiation at the surface for July from the MRI CGCM2.3.2a simulations. (b) The same as (a) but for January. The trend is defined as the 2080–2100 mean climatology minus the 1980–2000 climatology. Color scales with units of W/m^2 are shown on the right.

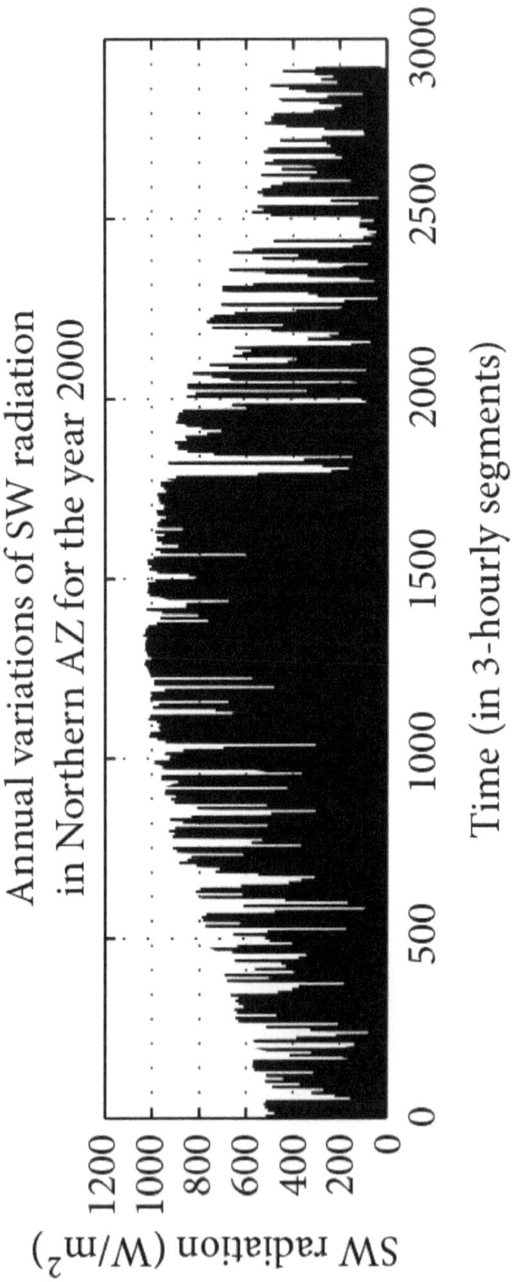

FIGURE 5: The time series of three-hourly downward shortwave radiation at surface, averaged over multiple grid points that cover Northern Arizona, for the year 2000 from the GFDL CM2.0 20th century simulation. The unit on the ordinate is W/m^2.

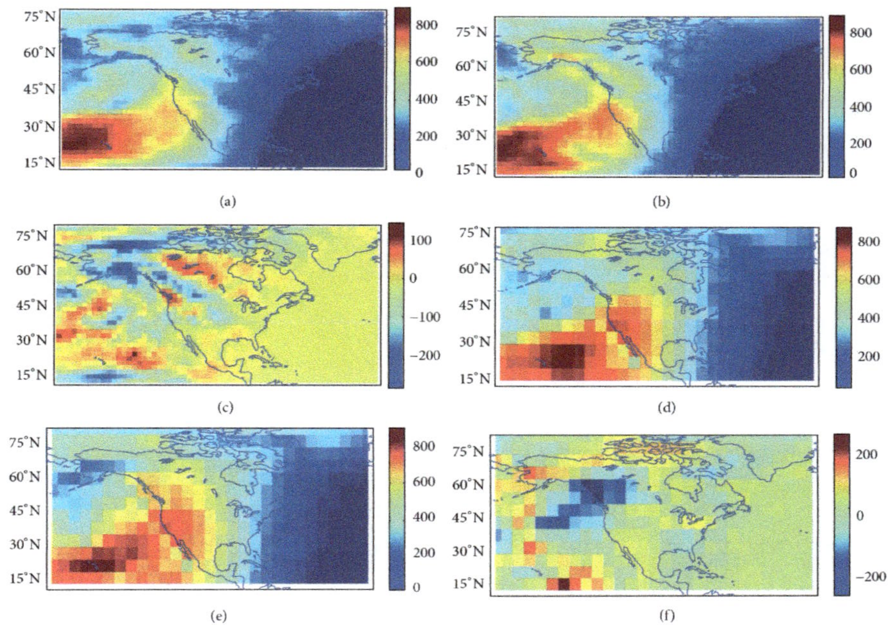

FIGURE 6: (a) The downward shortwave radiation at the surface at 3 PM local time of US West Coast, averaged over July 2100, from the GFDL CM2.0 21st century run. (b) The same as (a) but for the average over July 2000 from the GFDL CM2.0 20th century run. (c) The trend, defined as (a) minus (b). Panels (d)–(f) are the counterparts of (a)–(c) from the MRI CGCM2.3.2 simulations. The color scale with units of W/m^2 is shown on the right for each panel.

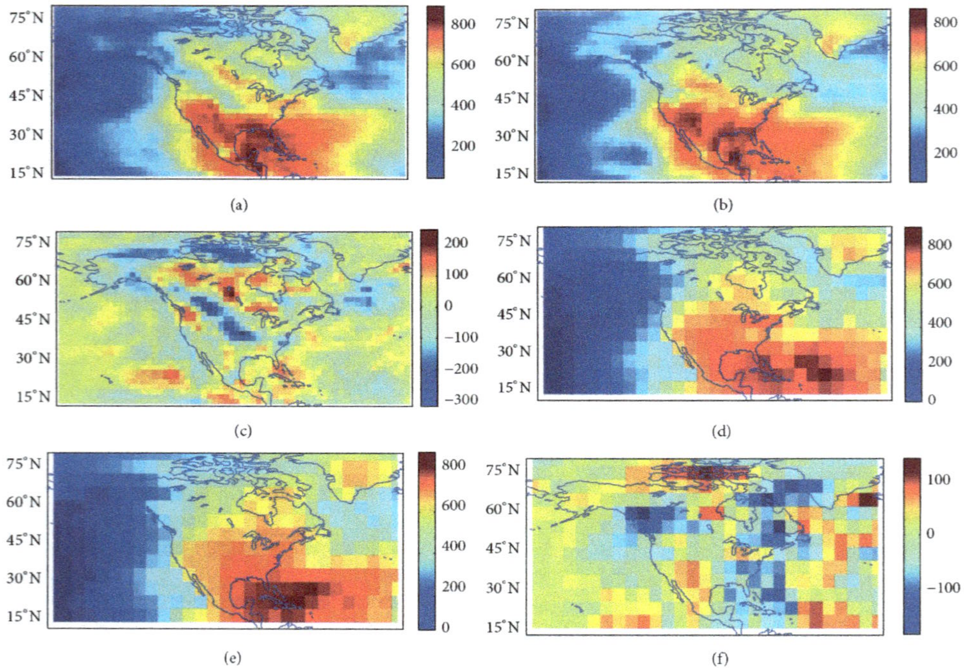

FIGURE 7: The same with Figure 6 but for the downward shortwave radiation at 9AM local time of US West Coast for July. Panels (a)–(c) are from GFDL CM2.0 and panels (d)–(f) are from MRI CGCM2.3.2 simulations. The color scale with units of W/m^2 is shown on the right for each panel.

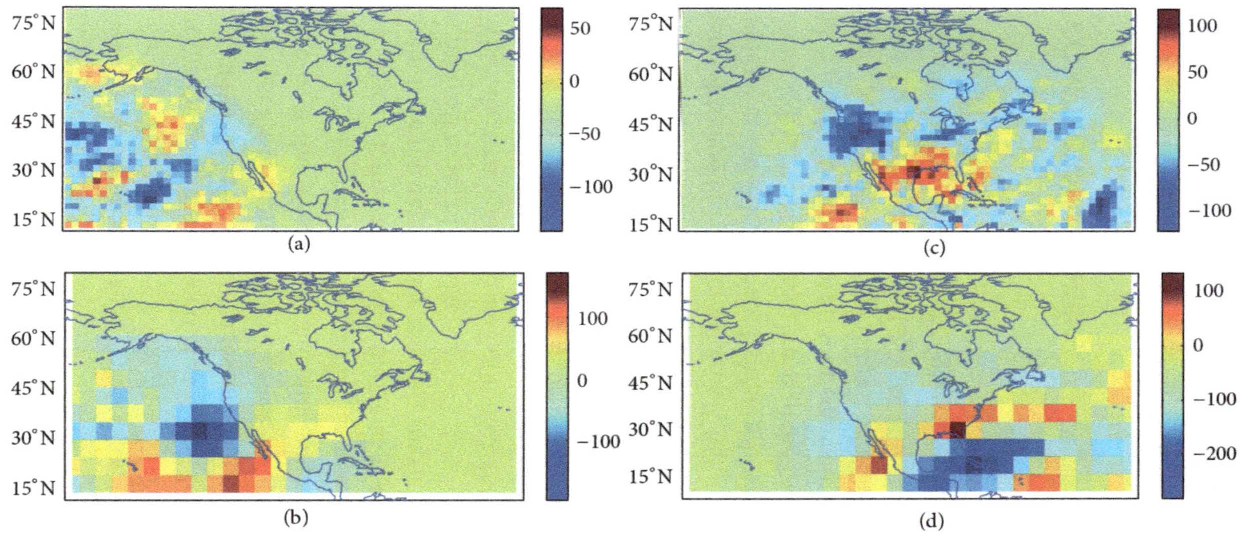

FIGURE 8: (a) The trend of downward shortwave radiation at surface at 3 p.m. local time of US West Coast, defined in the same manner as Figure 6(c) but for the average over January, from GFDL CM2.0 simulations. (b) The same as (a) but from MRI CGCM2.3.2 simulations. Panels (c) and (d) are the counterparts of (a) and (b) but for 9 a.m. local time of US West Coast, also averaged over January. The trends in all four panels are defined as January 2100 average minus January 2000 average. The color scale with units of W/m^2 is shown on the right for each panel.

Figures 6(a)–6(c) are similar to Figures 1(a)–1(c) but for the GFDL CM2.0 simulations of the shortwave radiation at 3 PM local time of US West Coast. Figure 6(a) is the average over July 2100 and Figure 6(b) the average over July 2000. The trend, defined by 2100 minus 2000, is shown in Figure 6(c). Figures 6(d)–6(f) are the counterparts of Figures 6(a)–6(c) but for the MRI CGCM2.3.2 simulations. Figure 7 is similar to Figure 6 but for the shortwave radiation at 9 a.m. local time of US West Coast. Note that the climatological values and trends in Figures 6 and 7 are much higher than the monthly mean values in Figure 1 because the latter are the average over the whole day, including nighttime. Just like the monthly mean, in July, the trends projected by the two models are significantly different. Also, within each model, the trend at a particular time of the day is different from the trend of the monthly mean which includes the contributions from all times of the day.

While the trends in July are not robust, the trends in January as shown in Figure 8 exhibit a greater degree of consistency between the two models and across different times of the day. For conciseness, only the trends are shown. Figures 8(a) and 8(b) are the January trend at 3 p.m., and Figures 8(c) and 8(d) are the January trend at 9 a.m. local time of US West Coast. The top row (Figures 8(a) and 8(c)) is from GFDL CM2.0 and bottom row (Figures 8(b) and 8(d)) from MRI CGCM2.3.2 simulations. The decrease in sunlight at the surface over Northern USA and the increase of it over Southern USA, as previously shown in the monthly mean plots in Figures 3 and 4, can be identified in all four panels in Figure 8. From Figures 8(c) and 8(d), at the time when the entire United States is in the middle of the day, the January trend of the decrease in solar radiation in Northern USA can locally reach 100 W/m^2, or slightly over 10% of the local climatological value at that time.

3.4 CONCLUDING REMARKS

The most robust finding of this study is the wintertime (January) trend in the downward shortwave radiation at the surface over the United States. It exhibits a simple pattern of a decrease of sunlight over Northern USA and an increase of sunlight over Southern USA. This structure is simulated by both GFDL and MRI models and can be identified even at different times of the day. It is broadly consistent with the known poleward shift of storm tracks in the cold season in climate model simulations under an increasing GHG forcing. The negative trend in Northern USA is more prominent. Quantitatively, the centennial trend of the downward shortwave radiation at the surface in that region is on the order of 10% of the climatological value for the monthly mean (averaged over all times of the day) and slightly over 10% at the time when it is midday in the United States. This indicates a nonnegligible influence of the GHG forcing on solar energy in the long term. Nevertheless, when dividing the 10% by a century, in the near term, the impact of the GHG forcing is relatively minor such that the estimate of solar power potential using present-day climatology will remain useful in the coming decades. The global climate models used in this study have relatively coarse resolutions with the horizontal grid size exceeding 100 km. For the assessment of solar power potential at a specific site of existing or future solar power plant, it will be desirable to perform climate downscaling (e.g., Mearns et al. [11] and Sharma and Huang [12]) to take into account the effects of small-scale topography on cloudiness. The findings of this work will serve as a useful reference for future studies in that direction.

REFERENCES

1. Z. Li, H. W. Barker, and L. Moreau, "The variable effect of clouds on atmospheric absorption of solar radiation," Nature, vol. 376, no. 6540, pp. 486–490, 1995.
2. E. Maxwell, R. George, and S. Wilcox, "A climatological solar radiation model," in Proceedings of the Annual Conference, American Solar Energy Society, p. 6, Albuquerque, NM, USA, 1998.
3. R. George and E. Maxwell, "High-resolution maps of solar collector performance using a climatological solar radiation model," in Proceedings of the 1999 Annual Conference, p. 6, American Solar Energy Society, Portland, Me, USA, 1999.
4. G. A. Meehl, C. Covey, K. E. Taylor, et al., "The WCRP CMIP3 multimodel dataset: A new era in climatic change research," Bulletin of the American Meteorological Society, vol. 88, no. 9, pp. 1383–1394, 2007.
5. R. Seager, M. Ting, I. Held et al., "Model projections of an imminent transition to a more arid climate in southwestern North America," Science, vol. 316, no. 5828, pp. 1181–1184, 2007.

6. N. C. Baker and H.-P. Huang, "A comparative study of precipitation and evaporation in semi-arid regions between the CMIP3 and CMIP5 climate model ensembles," Journal of Climate, vol. 27, pp. 3731–3749, 2014.
7. K. E. Taylor, R. J. Stouffer, and G. A. Meehl, "An overview of CMIP5 and the experiment design," Bulletin of the American Meteorological Society, vol. 93, no. 4, pp. 485–498, 2012.
8. M. Wild, "Short-wave and long-wave surface radiation budgets in GCMs: a review based on the IPCC-AR4/CMIP3 models," Tellus A, vol. 60, no. 5, pp. 932–945, 2008.
9. V. P. Meleshko, V. M. Kattsov, V. A. Govorkova, P. V. Sporyshev, I. M. Shkol'nik, and B. E. Shneerov, "Climate of Russia in the 21st century. Part 3. Future climate changes calculated with an ensemble of coupled atmosphere-ocean general circulation CMIP3 models," Russian Meteorology and Hydrology, vol. 33, no. 9, pp. 541–552, 2008.
10. J. H. Yin, "A consistent poleward shift of the storm tracks in simulations of 21st century climate," Geophysical Research Letters, vol. 32, no. 18, Article ID L18701, pp. 1–4, 2005.
11. L. O. Mearns, R. Arritt, S. Biner et al., "The north american regional climate change assessment program: overview of phase I results," Bulletin of the American Meteorological Society, vol. 93, no. 9, pp. 1337–1362, 2012.
12. A. Sharma and H.-P. Huang, "Regional climate simulation for Arizona: impact of resolution on precipitation," Advances in Meteorology, vol. 2012, Article ID 505726, 13 pages, 2012.

CHAPTER 4

Climate Change Impact on Photovoltaic Energy Output: The Case of Greece

IOANNA S. PANAGEA, IOANNIS K. TSANIS,
ARISTEIDIS G. KOUTROULIS, AND MANOLIS G. GRILLAKIS

4.1 INTRODUCTION

Solar photovoltaic systems have largely penetrated the global energy market and especially Europe. According to the European Photovoltaic Industry Association [1], photovoltaics (PV) for second consecutive year is the dominant new source of electricity production installed in Europe, where 55% of the global market of new connected to the grid capacity in 2012 with 17.2 GW is installed. PV systems consist of a competitive alternative for the decarbonization of Europe's energy sector, as they cover 2.6% of the electricity demand and 5.2% of the peak electricity demand. In 2012,

912 MW of PV was installed in Greece, increasing the PV contribution to 4% of the electricity demands.

Market trends show an expected decrease in the PV system prices from up to 2.31 €/W in 2012 in the residential segment to as low as 1.30 €/W in 2022 [2]. PV market in several countries as Greece is influenced by the political decisions and financial support. Nowadays because of the hard recession there is a decrease in large-scale PV projects installation. However, there is still potential of the Greek PV market to grow.

There is strong correlation between irradiation and temperature [3]. The downward irradiation that reaches the troposphere and the earth surface is absorbed by the atmospheric particles and the earth surface, respectively, emitting back long wave radiation (in the infrared spectrum) that increases the ambient heat and thus the temperature. However, the downward irradiation is largely affected by the cloud cover. Clouds affect the irradiation in three main ways. Firstly, they block a fraction of the direct downward irradiation (and thus affect negatively the direct radiation that reaches the earth surface) [4]. Secondly, they diffuse the already absorbed fraction of the irradiation to all directions, increasing the diffusive irradiation that reaches the earth surface. Thirdly, they block part of the long wave radiation that was supposed to be emitted from earth, back to space (greenhouse effect). The latter affects positively the near surface air temperature. However, the near surface air temperature is largely affected by the air mass temperature. To account for all these interactions simultaneously, regional climate models (RCMs) are used to produce estimations of surface radiation components and temperature [5].

The performance of PV systems is largely influenced by internal and external factors such as the structural features, visual loss, aging, radiation, shading, temperature, wind, pollution, and electrical losses [6–9]. Climate change will impact temperature and irradiance and therefore will alter the output capacity of PV systems [10]. PV systems present a negative linear relationship between the energy output and the temperature change [11], while the increase of solar radiation is proportional to the PV energy output.

The use of high spatial resolution RCMs has become more common over global circulation models (GCMs), which may not be precise enough to describe local climatic processes [12]. The main disadvantage of RCMs is that model projections have considerable uncertainties. The major sources of uncertainty in climate change research lie in the techniques used to force RCMs with boundary conditions, downscaling methods, and greenhouse gases emissions scenarios [13].

RCMs tend to simulate meteorological data with different statistical characteristics related to the observed-measured values. The time independent component of the error is the bias [14]. Studies [12] have shown that both GCMs and RCMs tend to overestimate the temperature in regions that present wet winters and dry summers and especially during the summer in south-eastern Europe [15]. The use of bias correction is thus required in order to adjust the climate models output according to the existing climate regime. Boberg and Christensen [12], Haerter et al. [14], Christensen et al. [15], and Terink et al. [16] emphasize the necessity of bias correction in order for the forced impact models to derive useful results in hydrology, water resources management, and other climate applications.

The correction methodology depends on the data type, the temporal and spatial resolution of data, and the time scale. In order to cope with the uncertainties related to the different possible detailed realizations of the climate system, ensembles of climate models output are used. Different GCMs can be used to quantify the uncertainty related to the different physical parameterizations of the large-scale land and atmosphere processes. Moreover, different RCMs can also be used to account for the uncertainties related to the representation of smaller scale processes, such as cloud microphysics or precipitation convection.

4.2 METHODOLOGY

4.2.1 BIAS CORRECTION

RCM temperature and irradiance outputs were corrected for their biases in mean and standard deviation for each calendar month, following the methodology presented in Haerter et al. [14]. The bias in mean is corrected by subtracting the differences found between observed and modeled values and a correction to the model data is performed to conform to the variability of the historical data. This procedure takes the sequence of anomalies and scales them consistently with the observed historical variability. In the case where data follow normal distribution the transfer function is linear and is of the form shown in the following equation:

$$\chi_{sc}^{cor} = \left(\chi_{mod}^{sc} - \overline{\chi_{mod}^{con}}\right) * \left(\frac{\sigma_{obs}^{con}}{\sigma_{mod}^{con}}\right) + \overline{\chi_{obs}^{con}} \tag{1}$$

where χ_{sc}^{cor} is the final adjusted time series, χ_{mod}^{sc} is the raw model predictions for the scenario period, χ_{obs}^{con} and χ_{mod}^{con} are the mean of observed and modeled data for the control period, respectively, and σ_{obs}^{con} and σ_{mod}^{con} are the standard deviations of observed and modeled data for the control period, respectively.

The final adjusted model time series exhibits the appropriate baseline mean and standard deviation with respect to the observed data.

4.2.2 ESTIMATION OF PV ENERGY OUTPUT UNDER VARIABLE CONDITIONS OF TEMPERATURE AND IRRADIANCE

In order to estimate the potential percentage change in PV output, the fractional change $\Delta P_{PV}/P_{PV}$ is calculated from the ratio between (2) and (3) taken from Crook et al. [10]. Consider the following:

$$\frac{\Delta P_{PV}}{\eta_{ref}} = -\Delta T G_{tot}\beta c_2 + \Delta G_{tot}\left(1 - \beta c_1 + \beta T_{ref} - 2\beta c_3 - T\beta c_2\right) - \Delta G_{tot}^2\beta c_3 - \Delta G_{tot}\Delta T\beta c_2 + \Delta G\gamma\log_{10}(G_{tot} + \Delta G_{tot}) + G_{tot}\gamma\log_{10}\left(\frac{G_{tot} + \Delta G_{tot}}{G_{tot}}\right) \tag{2}$$

$$\frac{P_{PV}}{\eta_{ref}} = G_{tot}\left(1 - \beta\left(c_1 + c_2 T + c_3 G_{tot} - T_{ref}\right) + \gamma\log_{10}G_{tot}\right) \tag{3}$$

where ΔP_{PV} is the change in photovoltaic power output, η_{ref} is the reference photovoltaic efficiency, ΔT is the change in temperature between the baseline and the scenario period, ΔG is the change in irradiance between the baseline and the scenario period, T is the daytime temperature for the baseline period, estimated by (4) as it can be found in Crook et al. [10], G_{tot} is the irradiance over the daylight for the actual cloud cover for the baseline period, calculated by (5) taken from Crook et al. [10], and T_{ref} is the reference temperature in which the performance of PV cell is estimated by the manufacturer. β is the temperature coefficient set by cell material and structure, γ is the irradiance coefficient set by cell material and structure, and c_1, c_2, and c_3 are coefficients which depend on details of the module and mounting that affect heat transfer from the cell. Consider the following:

$$T = \bar{T} + \frac{\overline{DTR}}{4} \tag{4}$$

where DTR is the diurnal range of the temperature (difference between minimum and maximum temperature) and T is the monthly average temperature. Consider

$$G_{tot} = \bar{G} + \frac{t_{24h}}{t_{daylength}} \tag{5}$$

where G is the monthly average irradiance and $t_{daylength}$ is the time of the daylight, calculated as monthly average, for all latitudes of the study site every 0.25°.

4.3 CASE STUDY AREA AND DATA USED

Monthly means of surface temperature and irradiance projections were obtained for five regional climate models of the ENSEMBLES (http://ensemblesrt3.dmi.dk/) database, under the special report on emissions scenarios (SRES) A1B emission scenario of the Intergovernmental Panel on Climate Change (IPCC). The main objectives of the ENSEMBLES project were to provide an ensemble prediction system based on the state of the art of high resolution global and regional earth system models developed in Europe. The produced simulations were validated by using high resolution gridded datasets for Europe to produce an objective probabilistic estimate of uncertainty in future climate at the seasonal to decadal and longer timescales. More information about the ENSEMBLES project can be found in UK Met Office page http://ensembles-eu.metoffice.com/ [21]. Detailed information about used RCMs temporal resolution is presented in Table 1. The processing of the climate data was performed in the ENSEMBLES RCMs spatial resolution which is 0.25 degrees.

TABLE 1: Name, institute, driving GCM, and transient experiment period of each ENSEMBLES RCM used.

Acronym	Institute	Driving GCM	Duration
C4IRCA3 [17]	SMHI, Sweden	HadCM3Q16	1951–2100
ETHZ-CLM [18]	ETHZ, Switzerland	HadCM3Q0	1951–2099
MPI-M-REMO [19]	MPI, Germany	ECHAM5-r3	1951–2100
SMHIRCA [17]	SMHI, Sweden	BCM	1961–2099
CNRM-RM5.1 [20]	CNRM, France	APREGE_RM5.1	1950–2100

The E-OBS [22] dataset provided observed minimum and maximum temperature data between 1950 and 2000. The SoDa database (http://www.soda-is.com) between 1985 and 2005 with spatial resolution of 20 km was used as an observational irradiance dataset. The SoDa server provides daily irradiation time series over Europe, Africa, and Atlantic Ocean, which is accessible on a free basis [23]. It supplies information of high quality, matching the actual needs of users, with improved time-space coverage and sampling [24]. The SoDa irradiance is satellite derived data of HelioClim-1 Daily Solar Irradiance v4.0 (HelioClim-1 Database of Daily Solar Irradiance v4.0 derived from satellite data, MINES ParisTech, Armines, France). HelioClim databases use the Heliosat-2

[25] method to process the Meteosat Images. The Heliosat method converts images acquired by meteorological geostationary satellites, such as Meteosat (Europe), GOES (USA) or GMS (Japan), into data and maps of solar radiation received at ground level. Mines ParisTech produced the method Heliosat-2 in November 2002, partly with the support of the European Commission (project SoDa Contract DG "INFSO" IST-1999-12245). The accuracy of the HelioClim-1 data is discussed in detail by Lefèvre et al. [26]. They assessed the accuracy of the HelioClim-1 data against ground measurements of the WMO radiometric network (55 sites in Europe and 35 in Africa) between 1994 and 1997. The RMS error was found to be 35 W/m^2 (17%) for daily mean irradiance and 25 W/m^2 (12%) for monthly mean irradiance. However, the bias of HelioClim against the observations was found to be in overall less than 1 W/m2 for the whole dataset.

The estimation of the change in P/V potential over Greece was conducted at NUTS2 spatial discretization level (NUTS: nomenclature of units for territorial statistics). Moreover, standard coefficients of monocrystalline silicon cells were considered for (2) and (3). For a monocrystalline silicon cell the coefficients are set according to Lasiner and Ang [27] as it is proposed in Crook et al. [10]. Thus, $\beta = 0.0045$, $\gamma = 0.1$, $c_1 = -3.75^{\circ}$ C, $c_2 = 1.14$, and $c_3 = 0.0175^{\circ}$ Cm^{2}W^{-1}. Typical value for the reference temperature is $T_{ref} = 25^{\circ}$ C.

4.4 RESULTS

4.4.1 CHANGE IN PV ENERGY OUTPUT

The change in PV energy output was estimated for two subsequent projection periods, 2011–2050 and 2061–2100. The periods 1950–2000 for temperature and 1985–2005 for irradiance were defined as control periods. Figure 1 presents the observed, the raw model ensemble mean, and the difference between them for temperature and irradiation. It was found that the RCM ensemble mean overestimates the mean temperature over Greek domain from 1°C to 3°C. This was also mentioned in Boberg and Christensen [12]. Similarly, RCM data tend to heavily overestimate the irradiance over central parts of Greece with the overestimation to be as high as from 40 W/m^2 to 50 W/m^2 in some parts. However, over the southernmost part of Greece, RCMs represent better the mean irradiation. The difference between the observed and model values stresses the need of adjusting the bias in both the examined parameters. Figure 2 shows the change in RCM simulated mean irradiance and temperature for the three considered 40-year periods, after the adjustment of the bias. The relative change in mean for the two projected periods is also presented in Figure 3 compared to the control period. It is shown that in the first projection period (2011–2050) the mean increase ranges between 1 and 1.5°C, while by 2061–2100 period the increase range reaches up to 3–3.5°C for the most parts of the study area. Regarding irradiance the average increase for the first projection period ranges between 2 and 3 W/m^2 while for the second period there is a further increase of 2–5 W/m^2. More specifically, the largest increase is expected over central Peloponnese and over Western Greece, in the regions of Epirus and Western Macedonia.

The change in PV energy output was then estimated for each model and projection period. Figure 4 illustrates the projected mean change in PV output derived from the ensemble of the RCMs. By 2050 the average PV energy output could increase up to 1%-2% in Western and Southwestern Greece, whereas for the regions of Attica and Thessaly a decrease of 1% is projected. However in the second period Western Greece and specifically the regions of Epirus and Peloponnese are projected to have an increase from 2% to 3%. For Thrace the respective increase is projected to be near 2%; in Northern Greece (Macedonia), Crete, and Aegean islands, the PV performance is expected to increases up to 1% while in the regions of Attica and Thessaly the projected performance decreases up from 1% to 2%.

Figure 5 illustrates the long term trend of the models ensemble relative projected change in PV energy output, for all administrative regions of Greece (NUTS 2 level) and for the entire study area. For all administrative regions, except Attica, a slight increasing trend in PV output is expected

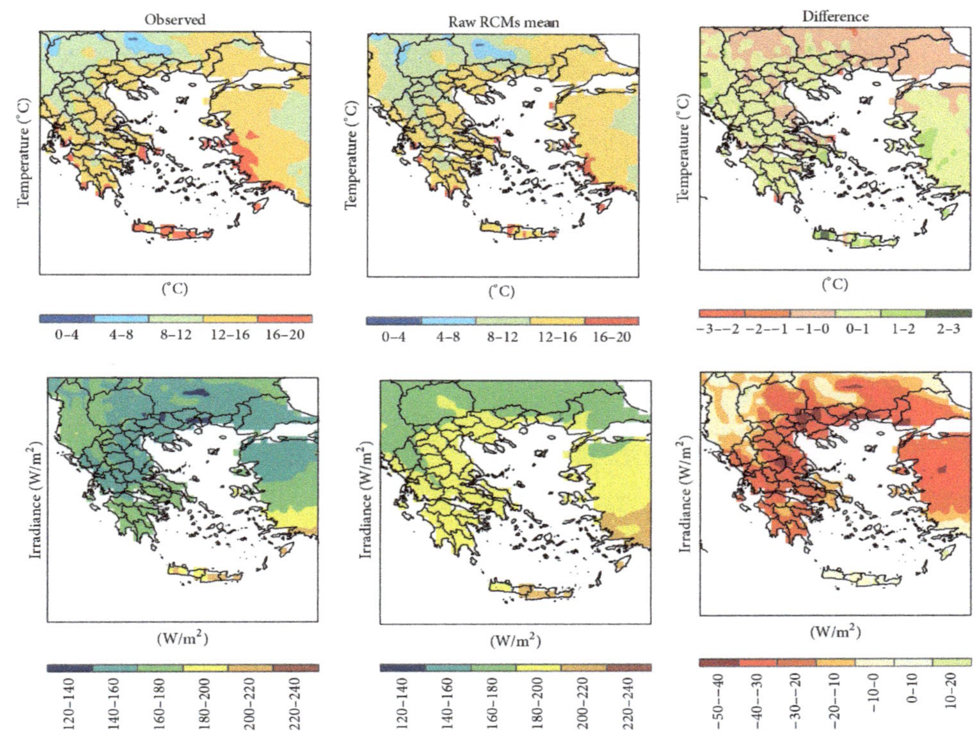

FIGURE 1: Maps of average temperature (upper) and irradiance (lower) for the control period based on observations (left panels), the ensemble mean (middle panels), and their difference (right panels).

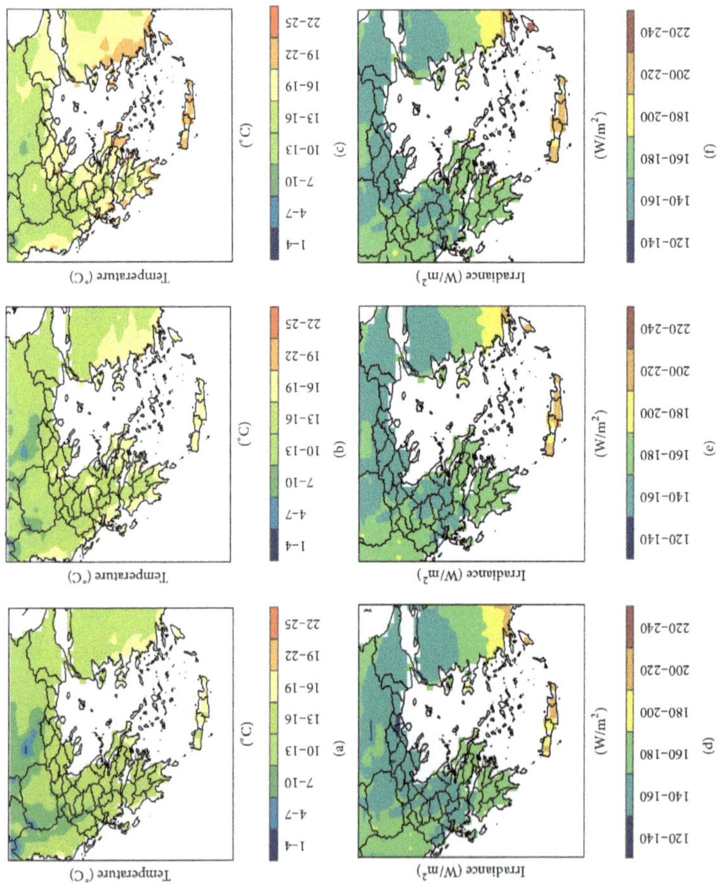

FIGURE 2: Mean temperature (upper) and irradiance (lower) fields over Greece for the reference (a, d) period, 2011–2050 period (b, e), and 2061–2100 period (c, f).

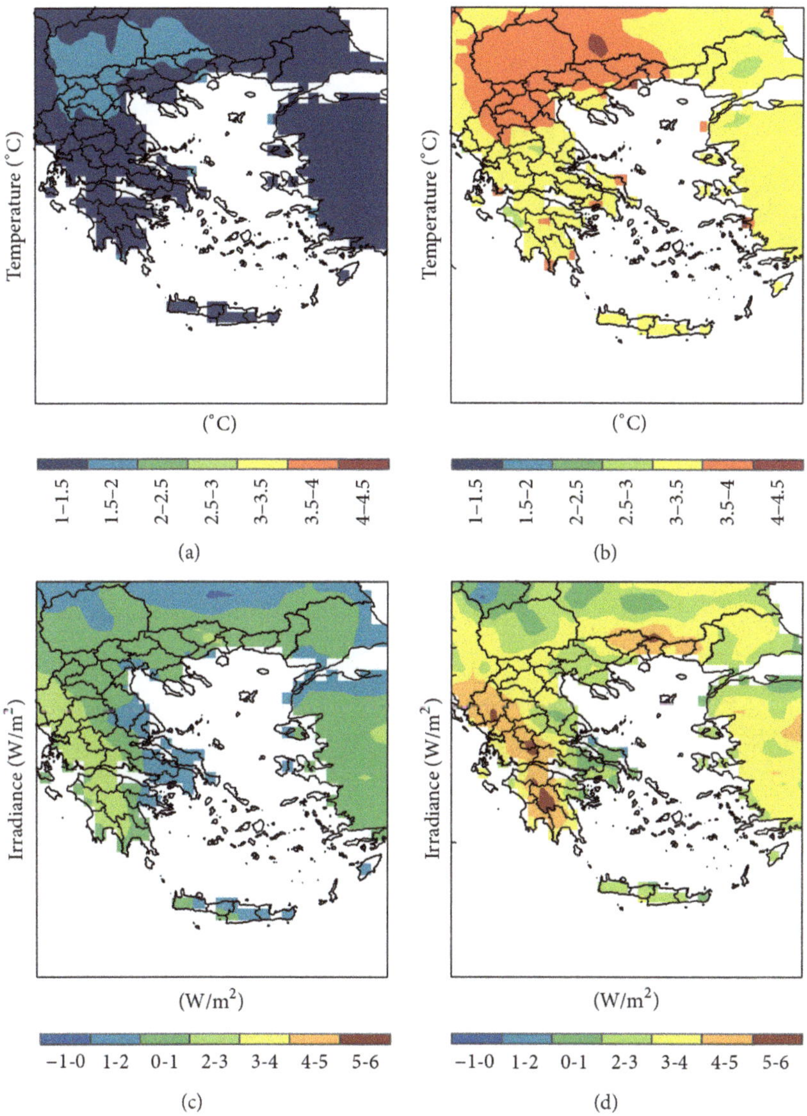

FIGURE 3: Absolute difference between reference and 2011–2050 period (a, c) and between reference and 2061–2100 (b, d) for temperature (upper) and irradiance (lower).

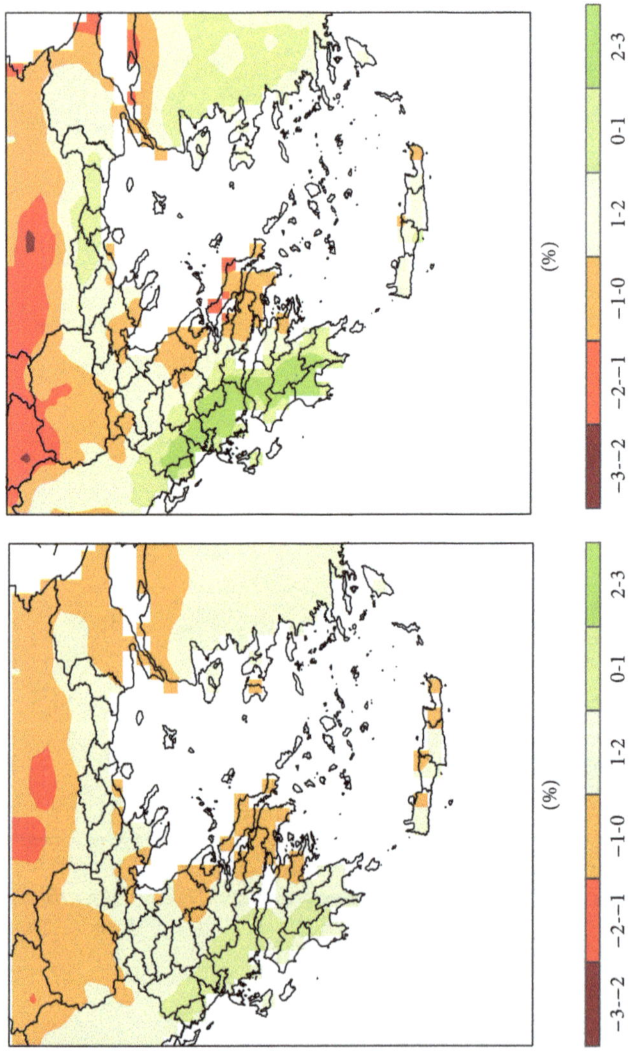

FIGURE 4: Estimated change in PV output for (a) the 2011–2050 period and (b) for the 2061–2100 period.

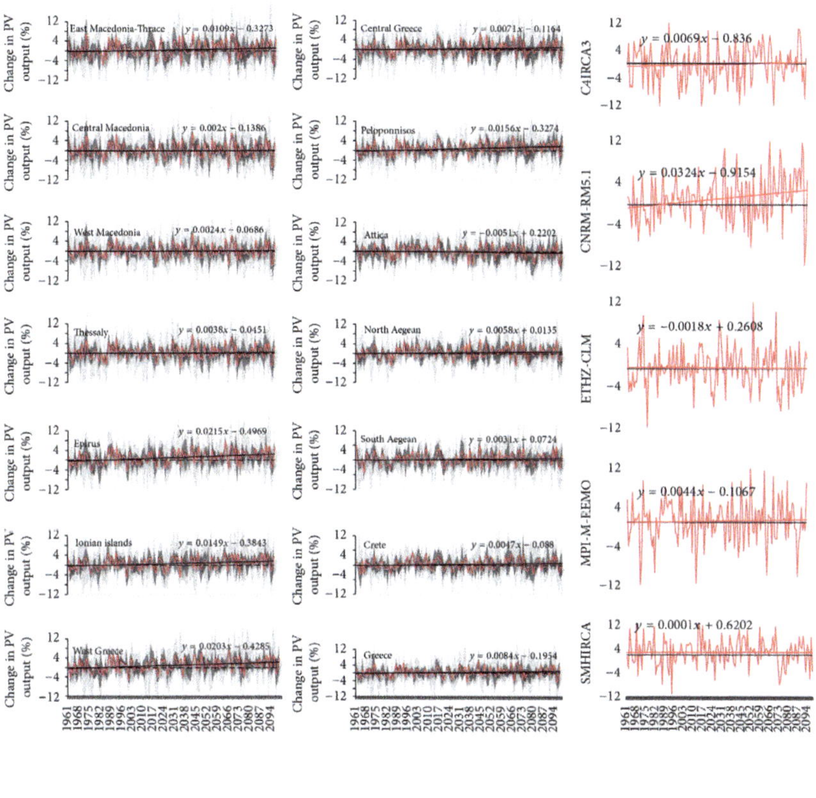

FIGURE 5: Relative change in PV output for the 13 prefectures of Greece and for the entire Greek region (left and center). Red line shows the average change of all models, dark grey represents the 50% range of the values of all RCMs, and light grey shows the 100% range of the models. Subplots in the right are the ensemble mean of each RCM for entire Greek region. The relative change is estimated using the 1960–2000 as baseline period.

during the 21st century ranging between 0.2% and 2% in the regions of Western Greece and Epirus.

It is important to note that in Figure 5, the range of relative change from all models values is between ±12%, which indicates the models' uncertainty. Referring to Crook et al. [10], the uncertainty in the specific research lies in the uncertainty of the projected changes in temperature and insolation. The most important source of uncertainty is the changes of cloud cover [28] and behavior as simulated in climate models, which plays an important role in climate change projections for the 21st century. For Southern Europe a decrease in cloud cover and increase in surface radiation are projected [10]. Therefore without the change in cloud cover the warming would still exist but not at this extend. Changes in cloud coverage influence the diurnal temperature variation [22] with greater influence on T_{max} during summer and T_{min} during winter than on the average daily temperature. This necessitates the distinctive consideration of T_{max} and T_{min} in climate change impact research. For Europe an increase in DTR (diurnal temperature range-temperature difference between maximum and minimum daily temperatures) is projected with greater uncertainty in the local summer season variation [29].

Figures 6 and 7 present the percentage change of PV performance as derived from each RCM for 2011–2050 and 2061–2100 periods, respectively, compared to the RCMs outputs for the reference period. Results indicate that the signal of the projected change in average PV performance over Greece is robust with large spatial variability, however, among the different RCMs. The majority of the RCMs project an increase between 1% and 2% over the most regions of the study area for the first period of study except from the RCM C4IRCA3 model which predicts a decrease up to 4% over the regions of Central and East Macedonia and the RCM MPI-M-REMO model which shows a decrease of about 2%. The projection for the second period indicates that, except for the RCM CNRM-RM 5.1 which shows a large increase, the rest of the RCM models predict an increase of the productivity of PV systems in the western mainland, Peloponnese, and Crete about 2%, while in

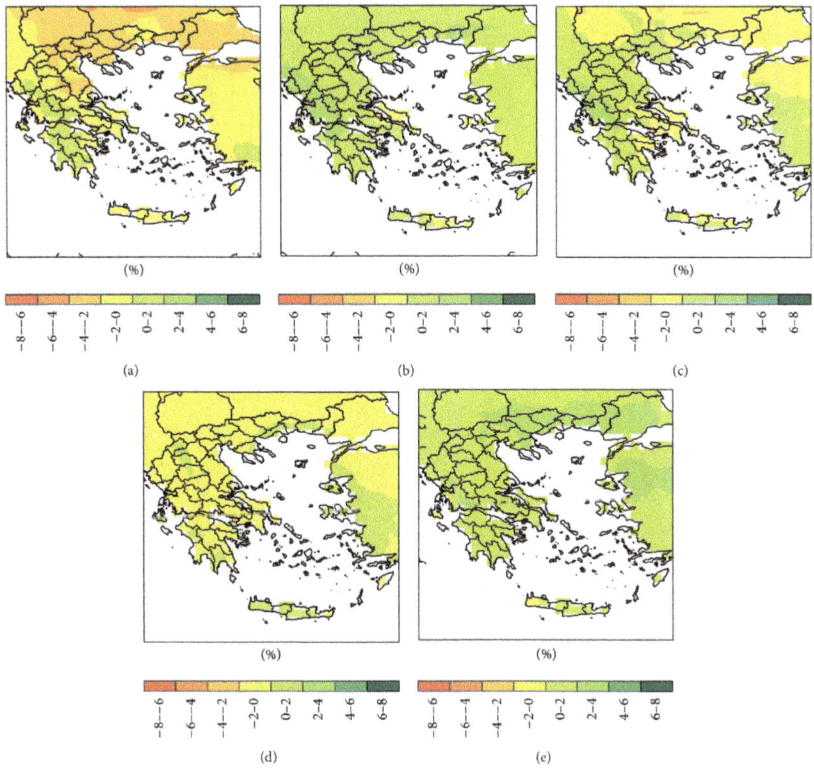

FIGURE 6: The first projection period (2011–2050) results in PV output change as it was estimated from each RCM ((a): C4I-RCA3, (b): CNRM–RM5.1, (c): ETHZ–CLM, (d): MPI-M-REMO, and (e): SMHI-RCA).

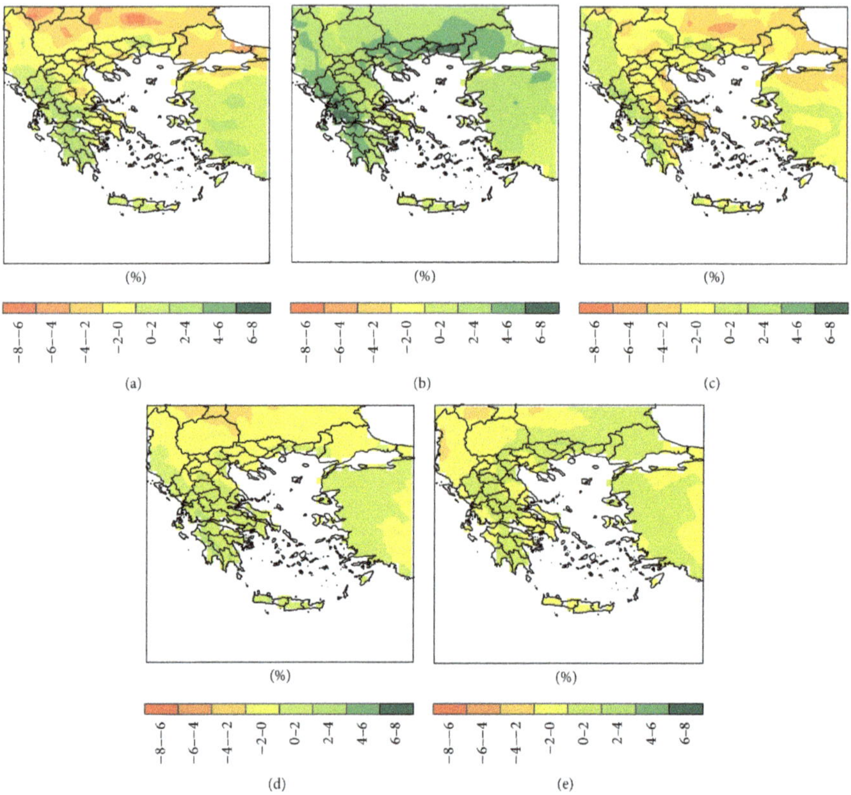

FIGURE 7: The second projection period (2061–2100) results in PV output change as it was estimated from each RCM ((a): C4I-RCA3, (b): CNRM–RM5.1, (c): ETHZ–CLM, (d) MPI-M REMO, and (e): SMHI-RCA).

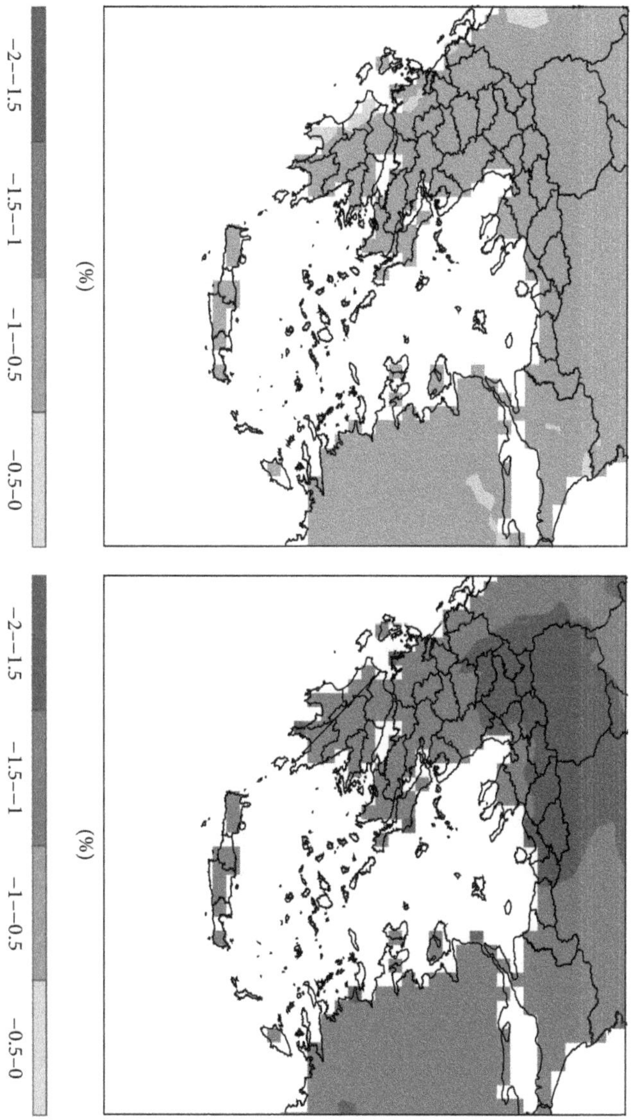

FIGURE 8: Temperature contribution to % PV output for 2011–2050 (a) and 2061–2100 (b) periods.

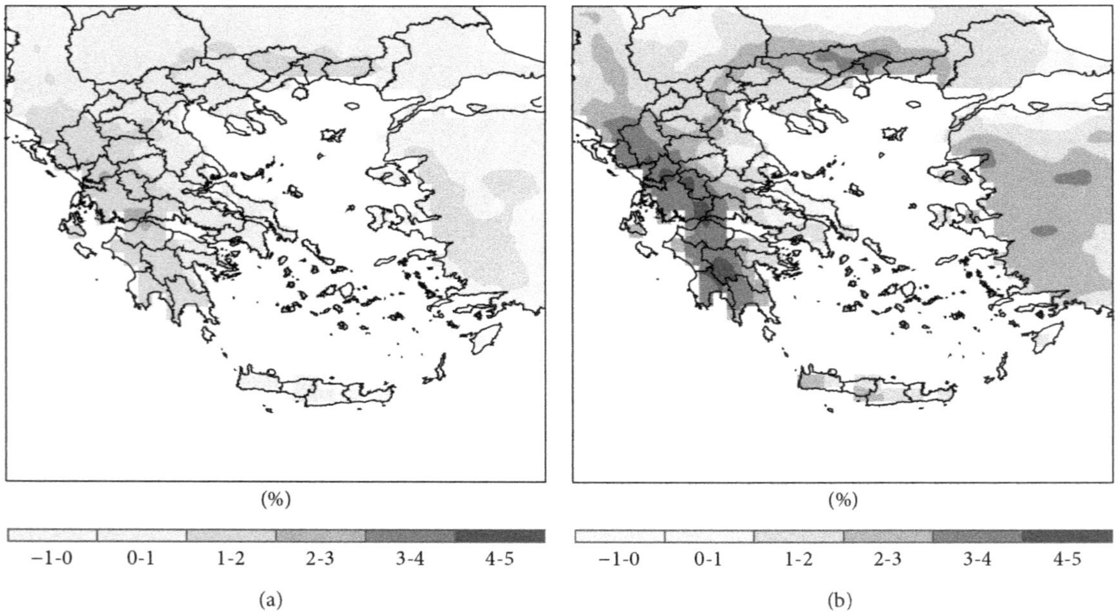

FIGURE 9: Irradiance contribution to % PV output for 2011–2050 (a) and 2061–2100 (b) periods.

the eastern mainland a decrease in productivity of about 2% and above is shown.

4.4.2 CONTRIBUTION OF TEMPERATURE AND INSOLATION IN THE PV OUTPUT CHANGE

The relative contributions of ΔT and ΔG are estimated by setting the projected change in irradiance and temperature, respectively, equal to zero in the calculations for the potential percentage change in the PV energy output in (2) following the methodology presented in Crook et al. [10]. In Figure 8, the ensemble mean PV output change is estimated for both projected periods, by setting the irradiance change equal to zero. In Figure 9 the respective results are presented by setting the temperature change equal to zero. The individual change in irradiance results in a significant increase on PV energy output up to 5% while the increase of temperature causes a decrease up to 2%. The relative contribution of each parameter varies from region to region. As it is expected, the two factors present different correlation signals with the PV output. In some cases the change in temperature and irradiance mutually compensates resulting in a small net change in performance. This is mostly the case in the first study period. However, there are cases such as the eastern part of Greece (Attica, Thessaly) where an increase in temperature of about 3°C–3.5°C cannot be compensated by the irradiation increase of about 1–3 W/m^2 and the estimation of PV energy output is negative, resulting in an overall reduction of PV output up to 3%. In contrast, over the Western Greece, despite the temperature increase at the same levels, the overall performance is expected to increase due to significant irradiation increase as a result of lower cloud coverage. Nonetheless, in some cases the large temperature increase cannot be compensated by the increase irradiance, as it can be observed in both study periods in eastern parts of the Greek mainland, some areas in Central Macedonia, and over Crete.

4.5 CONCLUSIONS

The projections of temperature and irradiance from 5 RCMs were analyzed for their contribution to monoclystalic PV panels' output change, over Greece. The analysis was conducted for two future periods, 2011–2050 and 2061–2100. The RCMs project an average temperature increase up to 1.5°C for the period 2011–2050 and from 3°C to 3.5°C for the period 2061–2100. Regarding the irradiance projections, an increase of 2-3 W/m^2 by 2011–2050 and up to 5 W/m^2 by 2061–2100 could be expected.

The PV output is projected to have an increasing trend for all regions of Greece during the 21st century. The region of Attica is an exemption where a reduction of 0.5% is projected. In the first period of study, an average increase between 1 and 2% in the majority of the study area is likely to occur except for the region of Attica and some areas in Thessaly and Central Macedonia. During the second period (2061–2100) a significant increase in the PV output is projected over the western mainland of Greece and Peloponnese, while a mild decrease for the eastern mainland and Central Macedonia is expected.

Examining the relative contributions of temperature and irradiance, a significant reduction due to the temperature increase is foreseen which is, however, outweighed by the irradiance increase, resulting in an overall increase in photovoltaic systems.

While the analysis shows clear increasing trends in the output of the monocrystalline PV systems over Greece, the rate of increase is small comparing to the variability amongst different RCMs. This is mainly attributed to the RCM shortcoming to robustly project the cloud cover and thus the irradiation parameter. The model's ability to capture the irradiation is expected to improve in the forthcoming Euro CORDEX [30].

REFERENCES

1. EPIA, Global Market Outlook for Photovoltaics 2013–2017, European Photovoltaic Industry Association, 2013.
2. EPIA, Connecting the Sun—Solar Photovoltaics on the Road to Large-Scale Grid Integration, European Photovoltaic Industry Association, 2012.
3. M. Wild, A. Ohmura, and K. Makowski, "Impact of global dimming and brightening on global warming," Geophysical Research Letters, vol. 34, no. 4, Article ID L04702, 2007.

4. D. Aiguo, K. E. Trenberth, and T. R. Karl, "Effects of clouds, soil moisture, precipitation, and water vapor on diurnal temperature range," Journal of Climate, vol. 12, no. 8, pp. 2451–2473, 1999.
5. R. F. Cahalan, G. Wen, J. W. Harder, and P. Pilewskie, "Temperature responses to spectral solar variability on decadal time scales," Geophysical Research Letters, vol. 37, no. 7, 2010.
6. M. Mani and R. Pillai, "Impact of dust on solar photovoltaic (PV) performance: research status, challenges and recommendations," Renewable and Sustainable Energy Reviews, vol. 14, no. 9, pp. 3124–3131, 2010.
7. M. E. Meral and F. Diner, "A review of the factors affecting operation and efficiency of photovoltaic based electricity generation systems," Renewable and Sustainable Energy Reviews, vol. 15, no. 5, pp. 2176–2184, 2011.
8. F. Dincer and M. E. Meral, "Critical factors that affecting efficiency of solar cells," Smart Grid and Renewable Energy, vol. 1, pp. 47–50, 2010.
9. E. Skoplaki and J. A. Palyvos, "On the temperature dependence of photovoltaic module electrical performance: A review of efficiency/power correlations," Solar Energy, vol. 83, no. 5, pp. 614–624, 2009.
10. J. A. Crook, L. A. Jones, P. M. Forster, and R. Crook, "Climate change impacts on future photovoltaic and concentrated solar power energy output," Energy and Environmental Science, vol. 4, no. 9, pp. 3101–3109, 2011.
11. G. Notton, C. Cristofari, M. Mattei, and P. Poggi, "Modelling of a double-glass photovoltaic module using finite differences," Applied Thermal Engineering, vol. 25, no. 17-18, pp. 2854–2877, 2005.
12. F. Boberg and J. H. Christensen, "Overestimation of Mediterranean summer temperature projections due to model deficiencies," Nature Climate Change, vol. 2, no. 6, pp. 433–436, 2012.
13. J. Chen, F. P. Brissette, and R. Leconte, "Uncertainty of downscaling method in quantifying the impact of climate change on hydrology," Journal of Hydrology, vol. 401, no. 3-4, pp. 190–202, 2011.
14. J. O. Haerter, S. Hagemann, C. Moseley, and C. Piani, "Climate model bias correction and the role of timescales," Hydrology and Earth System Sciences, vol. 15, no. 3, pp. 1065–1079, 2011.
15. J. H. Christensen, F. Boberg, O. B. Christensen, and P. Lucas-Picher, "On the need for bias correction of regional climate change projections of temperature and precipitation," Geophysical Research Letters, vol. 35, no. 20, Article ID L20709, 2008.
16. W. Terink, R. T. W. L. Hurkmans, P. J. J. F. Torfs, and R. Uijlenhoet, "Bias correction of temperature and precipitation data for regional climate model application to the Rhine basin," Hydrology and Earth System Sciences Discussions, vol. 6, no. 4, pp. 5377–5413, 2009.
17. E. Kjellstrom, L. Barring, S. Gollvik et al., "A 140-year simulation of European climate with the new version of the Rossby Centre regional atmospheric climate model (RCA3)," SMHI Reports Meteorology and Climatology, 108, SMHI, SE-60176 Norrkoping, Sweden, 2005.
18. U. Bohm, M. Kucken, W. Ahrens et al., "CLM—the climate version of LM: brief description and long-term applications," COSMO Newsletter, vol. 6, 2006.

19. D. Jacob, "A note to the simulation of the annual and inter-annual variability of the water budget over the Baltic Sea drainage basin," Meteorology and Atmospheric Physics, vol. 77, no. 1–4, pp. 61–73, 2001.
20. R. Radu, M. Déqué, and S. Somot, "Spectral nudging in a spectral regional climate model," Tellus A: Dynamic Meteorology and Oceanography, vol. 60, no. 5, pp. 898–910, 2008.
21. C. D. Hewitt and D. J. Griggs, "Ensembles-based predictions of climate changes and their impacts," Eos, vol. 85, no. 52, p. 566, 2004.
22. M. R. Haylock, N. Hofstra, A. M. G. Klein Tank, E. J. Klok, P. D. Jones, and M. New, "A European daily high-resolution gridded data set of surface temperature and precipitation for 1950–2006," Journal of Geophysical Research D: Atmospheres, vol. 113, no. 20, Article ID D20119, 2008.
23. S. Cros, D. Mayer, and L. Wald, "The availability of irradiation data," Report IEA-PVPS T2-04, International Energy Agency, Photovoltaic Power System Programme, Centre d' Energetique, Ecole des Mines de Paris/Armines, Antipolis, France, 2004.
24. B. Gschwind, L. Ménard, M. Albuisson, and L. Wald, "Converting a successful research project into a sustainable service: the case of the SoDa Web service," Environmental Modelling and Software, vol. 21, no. 11, pp. 1555–1561, 2006.
25. C. Rigollier, M. Lefèvre, and L. Wald, "The method Heliosat-2 for deriving shortwave solar radiation from satellite images," Solar Energy, vol. 77, no. 2, pp. 159–169, 2004.
26. M. Lefèvre, L. Wald, and L. Diabaté, "Using reduced data sets ISCCP-B2 from the Meteosat satellites to assess surface solar irradiance," Solar Energy, vol. 81, no. 2, pp. 240–253, 2007.
27. F. Lasiner and T. G. Ang, Photovoltaic Engineering Handbook, Adam Higler, Princeton, NJ, USA, 1990.
28. K. E. Trenberth and J. T. Fasullo, "Global warming due to increasing absorbed solar radiation," Geophysical Research Letters, vol. 36, Article ID L07706, 2009.
29. D. B. Lobell, C. Bonfils, and P. B. Duffy, "Climate change uncertainty for daily minimum and maximum temperatures: a model inter-comparison," Geophysical Research Letters, vol. 34, no. 5, 2007.
30. D. Jacob, J. Petersen, B. Eggert, et al., EURO-CORDEX: New High-Resolution Climate Change Projections for European Impact Research Regional Environmental Change, Springer, Berlin, Germany, 2013.

PART III

HYDROPOWER AND CLIMATE CHANGE

CHAPTER 5

Assessing Climate Change Impacts on Global Hydropower

BYMAN HAMUDUDU AND AANUND KILLINGTVEIT

5.1 INTRODUCTION

Climate change is one of the great challenges of the 21st century [1]. The International Energy Agency (IEA) report of 2011 projected that renewables based electricity generation would triple between 2008 and 2035 under the increasing-use-of-renewables scenario. Hydropower generation makes a substantial contribution to meeting today's increasing world electricity demands. The report adds that the share of renewables in global electricity generation increases from 19% to almost a third (nearly the same as coal). The primary increase is said to come from hydropower and wind but hydropower remains dominant over the projection period. It is projected that global hydropower generation might grow by nearly 75% from year 2008 to year 2050 under business-as-usual scenario but that it could grow by roughly 85% over the same period in a scenario with aggressive action to reduce greenhouse gas (GHG) emissions. However, even under this latter scenario, increased hydropower generation is projected to provide only about 2% of the total GHG emission reductions from the global electric power sector compared to business-as-usual by year 2050 (with all renewable technologies nonetheless providing nearly 33.5% of GHG abatement from the power sector). According to IEA, a realistic potential for global hydropower is 2 to 3 times higher than the current generation, with most remaining development potential existing in Africa, Asia, and Latin America. IEA also notes that, while run-of-river (smaller) hydropower plants could provide as much as 150 to 200 GW of new generating capacity worldwide, only 5% of the world's small-scale (i.e., small, low, and hydro) hydropower potential has been exploited [2].

In year 2009, hydropower accounted for about 16% (approximately 3551 TWh/a) of total global electricity generation and has reached 26% of the total installed capacity for electricity generation [3]. Global generation of hydropower has been growing steadily by about 2.3% per year on average since 1980 while the EU reports increases of up to 3.1% per year for the European Union. Global average growth rates of hydropower generation in the future are estimated to continue in the range of 2.4–3.6% per year between 1990 and 2030 (EIA, 2009). The highest growth rates are expected in developing countries which have high unexploited hydropower potentials, but also in other countries, for example, parts of Eastern Europe. In Western Europe, an annual increase of only 1% is estimated [4]. In contrast to the above, there are also indications that the annual energy generation of some existing hydropower stations in some parts of the world has decreased since the 1970s, for example in some parts of Europe [5]. The reductions have generally been attributed to changes in average discharge, but it is not clear whether they reflect cyclic fluctuations, steadily rising water abstractions for other uses, or the consequences of long-term changing climate conditions. Recent climate studies have pointed out that the time has come to move beyond the wait-and-see approach in future climate scenarios. Projections of changes in runoff are supported by the recently demonstrated climate models. The global pattern of observed annual stream flow trends is unlikely to have arisen from unforced variability and is consistent with modeled response to climate forcing [6].

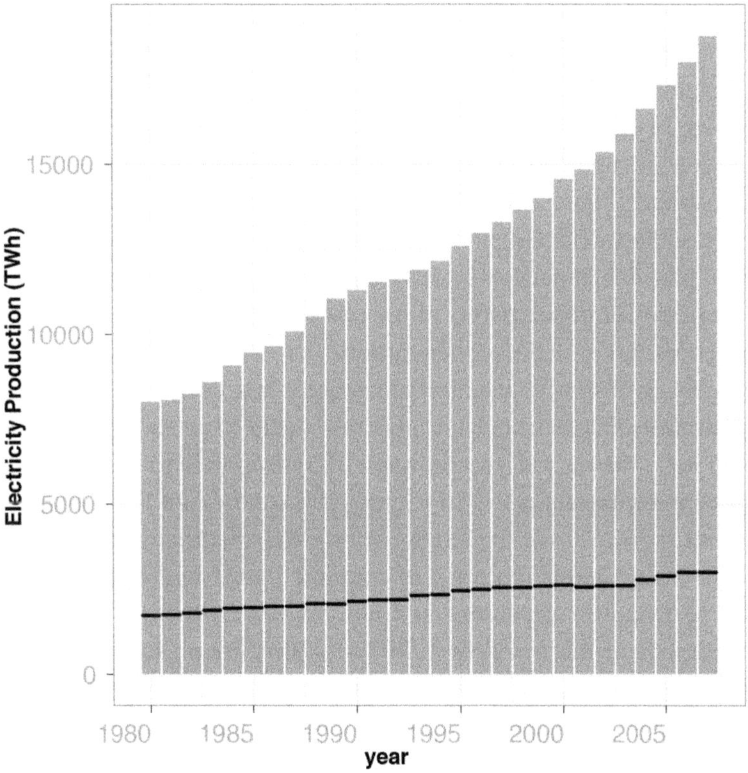

FIGURE 1: Global Total Electricity Generation Trends (TWh) in the last 20 years.

The IPCC in its AR4 concluded that climate change is occurring faster than earlier reported [7–9]. Many future climate scenarios point to the fact that the climate is changing rapidly although there are many arguments over the causes of these changes. Climate change will result in changes in various river flow conditions such as timing and quantity, sediment load, temperature, biological/ecosystem changes, and fish responses [10]. Climate change and the resulting changes in precipitation and temperature regimes will affect hydropower generation. It is reported that hydropower systems with less storage capacities are more vulnerable to climate change, as storage capacity provides more flexibility in operations. Although hydropower systems may benefit from more storage and generation capacity, expansion of such capacities may not be economically and environmentally justified. These changes would affect hydropower generation in all regions of the world. Given the significant role of hydropower, the assessment of possible impacts of climate changes on regional discharge regimes and hydropower generation is of interest and importance for management of water resources in power generation.

Global hydropower generation capacity has been increasing steadily over the last 30 years, and the past few years have shown an increased growth rate. Figure 1 shows the ratio of hydropower to the total electricity generation from year 1980 to year 2008. Although the ratio is reducing from 0.20 to 0.16, the Figure shows that hydropower generation is also increasing and is projected to continue increasing till year 2050. The global hydropower capacities and the contributions from various continents/regions of the world from 1980 to 2008 are presented in Figure 2. Europe, America, and Asia have sizable share of hydropower capacities. The installed capacity for Europe and Northern America, though large, has not been increasing much during this period while that in Southern/Central America and Asia/Oceania has greatly increased during this period as seen in Figure 2. However, the continental potentials are different, large in other regions like Africa. Table 1 shows regional hydropower characteristics in terms of hydropower in operation, total potential, under-construction, planned and countries with more than 50% of their total electricity demand supplied by hydropower.

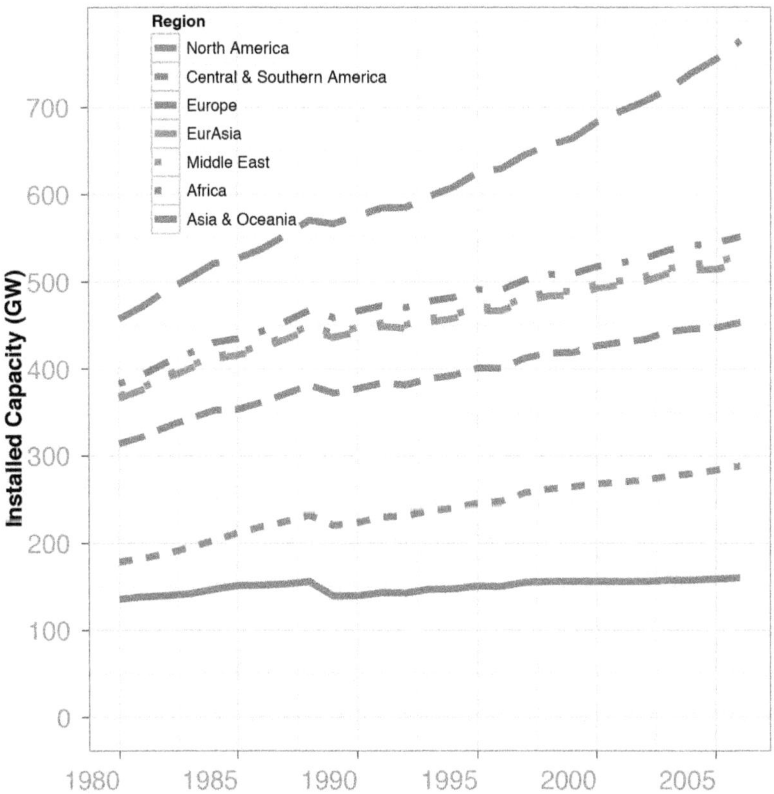

FIGURE 2: Trends in Global Installed Hydropower Capacities (1980–2006).

TABLE 1: World Hydropower in operation, under construction and Planned [3].

Region	Hydropower in Operation	% of Total Potential hydropower	Hydropower under construction	Hydropower Planned	Countries with 50% of electricity supply
	MW	%	MW	MW	#
Africa	23,482	9.3	5,222	76,600	23
Asia[1]	401,626	17.8	125,736	141,300	9
Europe[2]	179,152	53.9	3,028	11,400	8
North & Central America	169,105	34.3	7,798	17,400	6
South America	139,424	26.3	19,555	57,300	11
Australiasia/ Oceania	13,370	20.1	67	1500	4
World-Total	926,159		161,406	305,500	61

[1]Includes Russia and Turkey; [2]Excludes Russia and Turkey.

This study provides an overview of present (existing) global hydropower generation and its future prospects with respect to climate change. The focus of this work is global (all countries) i.e., low resolution (less detail), although for clarity's sake, some large countries like Australia, Brazil, Canada, China, India and USA had to be subdivided into provinces or states. Assessment of climate change impacts on hydropower can be done at various levels of detail with different methods. On a global scale, low resolution analysis is acceptable as detailed modeling may be costly and tedious. While recognizing the fact that climate change impacts hydropower in different ways—volume of flow, timings of flow, etc., the analysis has been confined to changes in mean flows (volume of flow). In addition, there is no estimate of the future hydropower development as doing so would require more detailed data (national development plans or trends) for each state and country. The study aims to answer questions related to national, regional and global hydropower generation and the expected increases or decreases in the same due to future changes in climate and water availability, and the extent of such changes. In order to answer the above, GIS analysis has been utilized to understand and visualize regional scenarios of hydropower generation. The analysis makes no attempt to analyze the impact of climate change on electricity demand, as it focuses on the side of generation. The GIS has been used here as a tool to merge and analyze different databases in order to gain insights into the anticipated changes. The database included data on world countries hydropower capacities, generation, global water resources, global runoff, dams, hydropower plants, etc. Table 2 shows regional hydropower statistics and of special interest is the installed capacity and hydropower generation in 2009. The table highlights the technically feasible, annual average potential, and feasible increase. The capacity factor of a power plant is the ratio of the actual output of a power plant over a period of time and its output if it had operated at full nameplate capacity the entire time. The lowest capacity factor is in Europe and clearly shows that hydropower in Europe is used mainly for peaking purposes than in the other regions [3].

TABLE 2. Regional Hydropower Potential (2009). The table highlights the technically feasible, annual average potential, annual generation capacity, and feasible increase [3].

Region	Technically Feasible Potential	Capacity Potential	Installed Capacity	2009 Generation	Capacity Factor	Feasible Capacity Increase
	TWh/y	MW	MW	TWh/y		%
Africa	1750	424,277	23,482	98	0.47	1925
Asia	6800	1,928,286	401,626	1514	0.4	670
Australasia/ Oceania	200	55,351	13,370	37	0.41	408
Europe	1140	352,804	179,152	542	0.37	214
North America	1510	360,397	169,105	689	0.48	225
Latin America	2968	596,185	139,424	671	0.57	464
Total/Average	14,368	3,722,930	776,760	3551	0.44	

There are many methods of assessing climate change impacts on hydropower generation systems. The use of a method depends on many factors such as the level of detail required, the geographical coverage, hydropower system description, and observation data availability. For example, the level of detail required for a global assessment differ from that needed for basin level assessments. Many studies have carried out assessment of hydropower generation in different parts of the world in various ways. Usually basin level assessment involves downscaling from GCMs through detailed hydrologic modeling and hydropower simulations, while on a regional level assessment, details begin to reduce. The methods can be seen as stepped analyses, where as the modeling begins to be complex, the detail and data requirements also do, beginning at the global scale down to small basin scale. Medellin-Azuara et al. [11] used downscaled hydrologic data in customized modeling scheme to assess the adaptability and adaptations of entire California's water supply system to dry climate warming. Madani and Lund [12] used an energy-based hydropower optimization model, avoiding the conventional modeling (simulation/optimization) methods, due to the large number of hydropower plants in California. The model used was developed for low-resolution, system-wide hydropower studies [12]. In a rather more detailed study of the Danube basin, development of hydropower was modeled using a special, coupled-physically-based hydrological model for three hydropower plants [13]. Another study on changes to whitewater recreation in California's Sierra Nevada used only elevation and runs as the predictors in identification, mapping and geomorphic classification to anticipate changes in runoff volume and timing from climate warming [14].

In a non-conventional approach, a method of modeling high elevation hydropower systems was developed and applied in California [15]. The method is energy-based and optimization was carried out on energy generation data on a monthly time scale and seasonal energy storage capacities. However there are some limitations as pointed out [15]. The method is a simplified approach where detailed hydropower data is unavailable. It is a simple approach for developing a good representation of an extensive hydropower system with little time or resources for policy and adaptation studies. Based on the results of some applications, the method is said to be skillful and useful for studying large hydropower systems when there is less details required. The developed method can be used for studying the effects of climate change on a large hydropower system [15]. In the above method, a large hydropower system (national or regional, large basin) can be modeled. However at the global scale, a more simplified approach is necessary not only to reduce on the complexities but due to lack of data for such a thorough detailed approach.

FIGURE 3. Future (2050) Runoff changes (%) based on 12 GCMs under A1B scenario.

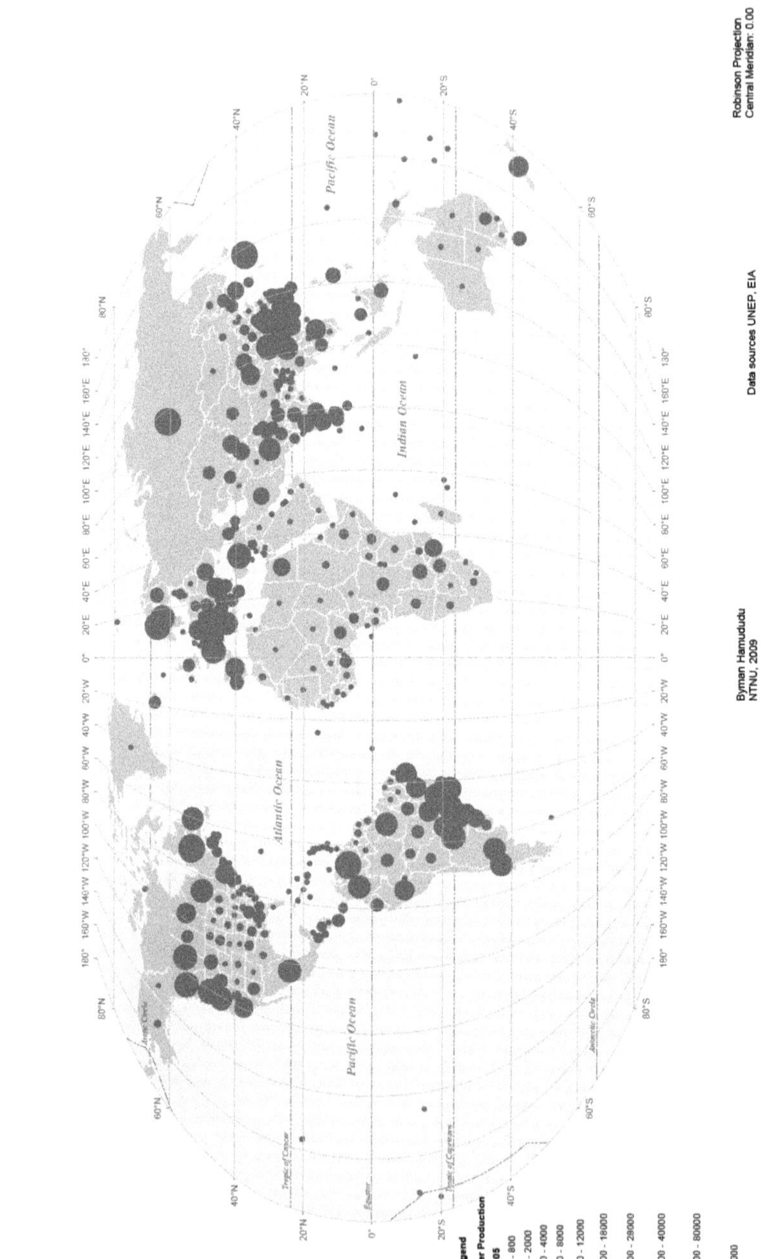

FIGURE 4: Hydropower generation (GWh) in 2005.

The approach used in this analysis aggregates different types of hydropower systems from different climates to highlight the larger global picture. The approach is based on the fact that the current hydropower generation system may only be limited by water availability. The main assumption is that if water supply reduces, the hydropower systems will likewise reduce generation and vice versa, assuming that current systems can be upgraded. With this approach, changes in annual mean flows are the main predictors of hydropower generation in each unit.

5.2 METHODOLOGY

The runoff baseline data is taken from the IPCC AR4 (2007), which is based on data supplied by Milly et al. [6]. An ensemble of 12 climate models was used with qualitative and statistically significant skill to simulate observed regional patterns of twentieth-century multi-decadal changes in streamflow. The realism of hydroclimatic simulations varies across models, so an ensemble from a subset of the models with the selection based on performance was used. The GCMs were ranked with respect to root-mean-square (r.m.s.) error (over the 165 basins and all runs) of the logarithm of long-term mean discharge per unit area; the logarithmic transform is commonly used in hydrology because flows can range over several orders of magnitude. A total of 12 GCMs were retained (35 runs of 20C3M) with the lowest error for use in the ensemble analyses [6]. Changes are expressed in terms of percentage variation from current runoff figures. The runoff changes are assessed at a national scale. On average, runoff can be thought of as the difference between the precipitation and evaporation over long periods of time and this makes it the available water for use, be it for hydropower, irrigation, domestic consumption, etc. In order to assess the future water availability, 12 GCMs with 20th century GRDC data [6] and future (A1B scenario) were used to evaluate the global trends of runoff. A total of 165 global basins with more than 28 years of data (greater 10% missing data) were used in regression analyses to predict the future resource availability. The model ensemble was in agreement in most regions, but there were some instances where the model ensembles did not produce similar trends and these were excluded from the analysis

[6]. The agreement criteria were based on 60% of the GCM agreeing on the trends of future runoff. In the countries where the GCM predictions did not agree, i.e., less than 66%, GCM having the same sign of increase or decrease were left out. The 12 GCMs results were tabulated and based on the above; a single value (median) was assigned to each country or state. The important measure of agreement was the trend, either positive or negative. The median was chosen as representing the mid-trend line of the GCMs for the particular unit, and so is not affected by the outliers. The mean was thus avoided, and the median was used in this analysis [6].

These estimated changes in runoff are the bases for country values (GCM estimates) and used as predictors in projecting hydropower generation for each country or state. The process data indicated that large changes in water resources can be expected in the coming decades due to climate changes across the globe. However, from this analysis, it is not possible to show the changes in seasons or in the timing of the water resources, which in some regions may be more pronounced. The changes are not weighted or did not have any spatial detail to represent the spatial variability in runoff areas within each country or state, and as such the results are generalized. The climate models do not simulate the high spatial resolution/detail in terms of projected climate change variables because of their large grid sizes. The runoff changes provided in this study are meant to provide a broad indication of the likely country based median changes.

Using GIS, the hydropower generation by countries were mapped into a GIS database system where different tables were merged for analysis. A GIS database management expedites the analysis on various tables that make up the database. The analysis was carried out on a national basis although some countries were subdivided into states due to their size; i.e., United States, Canada, Brazil, China, India and Australia. The countries or sub-regions were taken as units on which further analysis was based. The computed runoff changes is also mapped on a different layer. Computed future (2050) changes in runoff are based on results from 12 GCMs [6]. The GCMs differ in their future projections but a single value was sought by analyzing whether the general changes were positive or negative from most models. In all countries or states where the GCM agreed, in terms of trends, a median of the forecast of the GCMs was computed and the median value was then applied to annual hydropower generation for each of these units. The changes are then mapped to produce the future (year 2050) generation based on the current generation levels.

TABLE 3: GCMs used in the projections of future 2050 runoff changes after [5].

#	Model	Version	Modelling Centre	Country
1	CGHR	CGCM3.1 (T63),	Canadian Centre for Climate Modeling & Analysis	Canada
2	ECHOG		Meteorological Institute of the University of Bonn, Meteorological Research Institute of KMA, and Model and Data group,	Germany/ Korea
3	FGOALS	FGOALS-g1.0, LASG/	Institute of Atmospheric Physics,	China
4	GFCM20	GFDL-CM2.0	US Dept. of Commerce/NOAAA/Geophysical Fluid Dynamics Laboratory	USA
5	GFCM21	GFDL-CM2.1	US Dept. of Commerce/NOAAA/Geophysical Fluid Dynamics Laboratory	USA
6	GIEH	GISS-EH,	NASA Goddard Institute for Space Shuttles	USA
7	HADCM3	UKMO-HadCM3	Hadley Centre for Climate Prediction and Research/Met Office	UK
8	HADGEM	UKMO-HadGEM1	Hadley Centre for Climate Prediction and Research/Met Office	UK
9	MIHR	MIROC3.2 (hires),	Center for Climate System Research (The University of Tokyo), National Institute for Environmental Studies, and Frontier Research Center for Global Change (JAMSTEC)	Japan
10	MPEH5	MPEH5: ECHAM5/ MPI-OM,	Max Planck Institute for Meteorology	Germany
11	MRCGCM	MRI-CGCM2.3.2	Meteorological Research Institute	Japan
12	NCCCSM	CCSM3	National Center for Atmospheric Research	USA

Based on the above data, the analysis was carried out to convert changes in water resource availability to changes in hydropower generation. The runoff was assumed to be the main determinant of or limitation to hydropower generation. Results are given in the next section. The computational details are illustrated by a more detailed table for Africa (Table A1 where the database and computations can be seen for individual countries. The same level of detail has been applied for all other countries and sub-regions. The methodology is based on the fact that hydropower generation (N) is a function of flow (Q, in m^3 s^{-1}), head (H, in m) and efficiencies. The most varying factor is the flow (Q), referred to as water resources for every unit.

$$N = 9.81QH_{\eta} \tag{1}$$

The procedure uses the flow (Q) for the water resources for each country and assumes that the changes in water resources for that unit will impact the hydropower produced in the future. It is further assumed that most of the new hydropower developments will take place in the same regions where the existing systems are located. The results are expressed in percentage change relative to the generation of the existing system. This same percentage change is likely to occur even when the generating capacity is increased. Figure 4 shows data on hydropower generation; the sizes are proportional to the hydropower production for that country or state in year 2005.

5.3 DATA

Data were obtained from various sources and transformed where necessary into GIS layers. Most of the data of hydropower and energy were obtained from Energy Information Administration (EIA) of US, which is the official energy statistics of the US government freely available from their website [16] (Department of Energy 2009). Other national-level energy data were obtained directly from national websites and integrated into one database. GIS-related data like political boundaries and maps were obtained from UNEP geodata portal [17] (UNEP/DEWA/GRID-Europe, 2006), the data on dams from International Commission on large dams (ICOLD), national-level water resources data from Food and Agriculture Organization (Water Development and Management Unit, FAO) [18]. Data for trends and projections are based on a global runoff analysis by Milly (2005). Milly et al. showed global pattern of trends in stream flow and water availability in a changing climate. The study highlighted the variations in changes in runoff over the entire globe from region to region [5]. The following GCMs in Table 3 were used in the analysis. Runoff increases are predicted for the mainly northern regions of America, Canada, Europe and Russia as well as parts of India and Bangladesh, East Africa and a few countries in Southern America. The rest have reductions while for much of Central and West Africa, forecast cannot be made with certainty.

Table 4 shows the regions of the world and the countries grouped according to UNEP (2009). Note that some countries are unconventionally placed in regions, for example Russia and Turkey are grouped along with other Asia countries and not Europe. This changes the regional statistics i.e., adding the generation from Russia and Turkey to the already high hydropower production in Asia.

5.4 RESULTS AND DISCUSSION

The results from the analysis are shown in Figure 5. The size of the dots indicates the installed capacity while the colour (red for reduction and blue for increase) indicate the changes for each country/state where GCM prediction on runoff data were consistent and reliable (in agreement). Most of the highlights are in line with many site-specific studies on hydropower and climate in most of the regions of the world. The regions of Europe, US and Canada all have projections similar to results obtained in the studies [19–26].

Table 5 shows that 2931 TWh of hydro-electricity were produced in year 2005. From the analysis, based on 2005 global hydropower generation, it can be said that by year 2050, the hydropower generation would be affected differently in various regions of the world. There are regions where hydropower generation will increase and there are also regions where hydropower generation will decrease.

TABLE 4: Global Regional Groupings of the Countries according to UNEP(2009), after [17].

Continent	Region	Countries within the Region
Africa	Eastern	Burundi, Comoros, Djibouti, Ethiopia, Kenya, Madagascar, Mauritius, Reunion, Rwanda, Seychelles, Somalia, Tanzania, Uganda,
	Central	Central African Rep, Cameroon, Chad, Congo, Eq. Guinea, Gabon, Sao tome
	Northern	Algeria, Egypt, Libya, Morocco, Sudan, Tunisia, W. Sahara
	Southern	Angola, Botswana, Lesotho, Malawi, Mozambique, Namibia, South Africa, Swaziland, Zambia, Zimbabwe
	Western	Benin, Burkina Faso, Cape Verde, Gambia, Ghana, Guinea, Guinea-Bissau, Ivory coast., Liberia, Mali, Mauritania, Niger, Nigeria, Senegal, Sierra Leone, Togo
Asia	Central	Kazakhstan, Kirgizia, Tadzhikstan, Turkmenistan, Uzbekistan, Russia
	Eastern	China, Hong Kong, Japan, North Korea, South Korea, Mongolia, Taiwan
	South Eastern	Papua New Guinea, Brunei, Burma, Indonesia, Kampuchea, Laos, Malaysia, Philippines, Singapore, Thailand, Vietnam
	Southern west	Afghanistan, Bangladesh, Bhutan, India, Maldives, Nepal, Pakistan, Sri Lanka
	Western	Armenia, Azerbaijan, Bahrain, Cyprus, Georgia, Iran, Iraq, Israel, Jordan, Kuwait, Lebanon, Oman, Qatar, Saudi Arab, Syria, Turkey, United Arab Emirates Yemen
Australasia		Australia, New Zealand
Europe	Eastern	Belarus, Bulgaria, Czech republic, Estonia, Hungary, Latvia, Lithuania, Moldavia, Poland, Romania, Slovakia, Ukraine
	Northern	Denmark, Faroe island ., Finland, Iceland, Ireland, Norway, Sweden
	Southern	Albania, Bosnia and Herzegovina, Croatia, Greece, Italy, Macedonia, Malta, Portugal, San Marino, Serbia, Slovenia, Spain
	Western	UK., Austria, Belgium, France, Germany, Liechtenstein, Luxembourg, Netherlands, Switzerland
America	Caribbean	Anguilla, Antigua & b, Bahamas, Barbados, Cuba, Dominica, Domrep, Grenada, Guadalupe, Haiti, Jamaica, Martinique, Nantilles, Puerto Rico, St Chrs-nv, St Lucia, Stvinc & gr, Trinidad & Tobago, Turks & c.i,
	Central	Belize, Costa Rica, El Salvador, Guatemala, Honduras, Mexico, Nicaragua, Panama, Bermuda,
	Northern	Canada, USA
	Southern	Argentina, Bolivia, Brazil, Chile, Colombia, Ecuador, Falkland, French Guiana, Guyana, Paraguay, Peru, Surinam, Uruguay, Venezuela
Oceania		New .Caledonia, Solomon, Vanuatu, Cooking island, Guam, Kiribati, Nauru, Tuvalu, Fiji, French Polynes, Tonga, Hawaii, West Samoa

FIGURE 5: Percentage Changes in Global Hydropower generation resulting from 12 GCMs (AR4 2007) under A1B scenario.

In Africa, there are some countries with increasing hydropower generation and others with decreasing hydropower generation, as illustrated in the appendix. The Eastern African region shows increases in almost all countries except Ethiopia where there were disagreements among the GCMs. The Southern and Northern regions show decreases in hydropower generation. The Western region remains nearly the same but there are some countries with increases while others have decreases, and again here in most countries there were disagreements among GCMs on future runoff.

TABLE 5: Summary of Regional (2050) Changes in Hydropower generation.

Continent	Region	Generation TWh	Change TWh	% Change of total
Africa	Eastern	10.97	0.11	0.59
	Central	12.45	0.04	0.22
	Northern	15.84	−0.08	−0.48
	Southern	34.32	−0.07	−0.83
	Western	16.03	0.00	0.03
		89.60	0	−0.05
Asia[1]	Central	217.34	2.29	2.58
	Eastern	482.32	0.71	0.08
	South Eastern	57.22	0.63	1.08
	Southern	141.54	0.70	0.41
	Western	70.99	−1.66	−1.43
		996.12	2.66	0.27
Australasia/Oceania		39.8	−0.03	0
Europe[2]	Eastern	50.50	−0.60	−1.00
	Northern	227.72	3.32	1.46
	Southern	96.60	−1.79	−1.82
	Western	142.39	−1.73	−1.28
		517.21	−0.8	−0.16
America	Northern, Central/Caribbean	654.7	0.33	0.05
	Southern	660.81	0.30	0.03
		1,315.5	0.63	0.05
Global	2,931	2.46	0.08	

[1]Includes Russia and Turkey; 2 Excludes Russia and Turkey.

For Asia, positive trends owing to climate change have been projected for most countries. An exception is the Middle East (here grouped under Asia) which has decreasing trends. This continent shows the largest increases vis-a-vis the others. In fact, all the parts of this continent show increases apart from western part, which does not produce a lot of hydropower.

The Americas have a continental net increase with major producers having increases (south and north) and only central America having a reduced generation in the future. The northern part of America shows (mostly) increases and this changes southward with the central region of America showing decreases. Changes in the America nearly cancel out as decreases in some parts are offset by increases in others.

Southern, Eastern and Western Europe have reductions while the Northern part shows increased generation, and with increased generation in high-producing regions, the regional net growth is positive. The large producers are in the Northern region, and as such, the continental changes show net increases in hydropower generation.

Most of Australasia has reduced generation while Oceania shows an increase. There are disagreements among the GCMs on future projections over Australia. There are only a few states where there are agreements. This makes it difficult to make a good picture of future hydropower generation of this region.

From the results, it can be seen that most of the high hydropower-producing countries in the north (Canada, US and parts of Europe and Russia) will have increased generation, while for most of the south, whether big or small, hydropower generation will decrease.

It should be stated here that the analysis was carried out on a national basis (states for the largest countries), while this papers summarizes the results at a regional level. There are many differences within each region. Even when the overall region may register an increase, it is likely that some countries within the region may experience reductions. Table 5 has been appended to show intra-regional variations for one continent, Africa. Africa has been chosen to highlight these internal differences in changes due to its high hydropower potentials (undeveloped) and its having the greatest variations and the highest necessity for development in the future due to increasing population.

The global change in future hydropower generation due to climate change shows a slight increase over the current global hydropower generation (0.46 TWh). This could be improved by bringing on-stream fresh capacity either already under construction or on the anvil.

5.5 LIMITATIONS

The overall objective of this study was to present a global picture of impacts of climate change on hydropower generation. In order to do this efficiently, a lot of simplifications were made. These included ignoring the impacts such as changes in timing of flow, changes in sediment transport, etc. These are important factors in hydropower operation, but were not included in this analysis. In addition there were no adaption and/or mitigation on operations included in the analysis, and as such, no storage analysis or non-storage analysis was performed.

The changes are computed on the current hydropower generation and no future hydropower development has been included, firstly due to the fact that these data are difficult to obtain for each country or state for the whole world, and secondly because the analysis would become more complex, requiring more resources.

Another simplification is that changes are computed at country level (except for very large countries). The study recognizes that climate change impacts can vary spatially and sometimes over short distances, but again, the simplification that for each country, an average change is assumed may seem acceptable. The objective was to show the bigger global picture and the direction of change on the global scale.

The amount of electricity produced by a hydropower system depends on: (1) the discharge/flow (amount of water passing through the turbine per unit time); (2) the site head (the height of the water source); and (3) the turbine generating capacity and efficiency. In order to evaluate the impacts of climate change on hydropower globally, only the mean discharge/flow has been used as a factor to hydropower generation, which is also a simplification.

The above simplification would lead to some differences when the results presented in this study are compared to a more local detailed analysis of climate change impact on one or two hydropower system, where more plant data, time series data and detailed down-scaling is carried out. However a few comparisons made so far showed that the results were not very different (within ranges).

There are many factors that could be used to mitigate impacts on climate change on hydropower especially in operations. These have not been dealt with in this current study. Such factors include the storage capacity, pumped storage system, operation rule curve changes, etc. These were considered to be outside the scope of this study.

The primary function of a hydropower system is to generate power. However in many countries, the hydropower systems play important roles as general purpose water handling facilities. The multipurpose use of water and demand is important as the impacts of climate threaten the agreements that exist between many users of water. In areas projected with decrease, as the water resources decrease, competition and re-examinations of agreements may result. This ultimately would result in changes in the hydropower generation.

This study has not examined the impact of increased frequency of droughts and floods, as forecast in many places with climate change. If droughts and floods become more frequent, this scenario would severely impacts hydropower production. These extreme events would reduce the reliability of hydropower system to produce power. In regions where mean annual flow does not change, it is still possible that hydropower production would be severely affected if the droughts become more frequent. The impacts of changes in extreme events should be examined carefully on a local scale.

5.6 CONCLUSIONS

Hydropower generation is mainly influenced by runoff although there are other limiting factors. Changes in runoff will therefore lead to changes in hydropower generation. In its most accurate form, hydropower-plant based analysis for individual stations gives a better picture of future generation. However, when one is considering the global level, scale becomes an important issue.

The overall impacts on the global technical potential is expected to be slightly positive. However, results also indicate the possibility of substantial variations across regions and even within countries. Globally, hydropower generation computations show a very slight increase around year 2050 of about 0.46 TWh per annum. However, different countries and regions of the world will have significant changes; some with positive and others with negative changes. This study therefore provides general estimates of regional and global perspectives of the probable future hydropower generation scenarios.

Climate change is a challenge for the entire hydropower sector; the challenge is to come up with mitigation measures for hydropower operations and designs against these effects. Some regions have minimal infrastructure to act as a buffer the impacts of change.

The hydropower sector is one of the sectors least adversely affected on a global scale. Although the various regions will have varying changes, at the global level, there could be a slight gain in total global hydropower generation. It is worth mentioning here that after factoring in the uncertainty through the whole analysis process, it can be said that hydropower generation will remain nearly the same for some time into the future—till year 2050.

Investment (construction of new plants) in the hydropower sector could help reduce the gap (deficit) that may be created by effects of climate change on power generation in areas where there is still untapped potential. In other areas where the potential is nearly exhausted, better technology (e.g., high efficiencies) on existing systems would help mitigate the impacts or boost the contribution of hydropower to global electricity generation.

REFERENCES

1. IPCC. Special Report on Renewable Energy Sources and Climate Change Mitigation; Technical Report; Intergovernmental Panel on Climate Change: Geneva, Belgium, 2011.
2. EIA. International Energy Outlook 2011; U.S. Department of Energy, Energy Information Administration: Washington, DC, USA, 2011.

3. Bartle, A. Hydropower and Dams, World Atlas; Aqua Media International Ltd.: Sutton, UK, 2010.
4. Lehner, B.; Czisch, G.; Vassolo, S. The impact of global change on the hydropower potential of Europe: A model-based analysis. Energy Policy 2005, 33, 839–855.
5. Milly, P.C.D.; Betancourt, J.; Falkenmark, M.; Hirsch, R.M.; Kundzewicz, Z.W.; Lettenmaier, D.P.; Stouffer, R.J. Climate change: Stationarity is dead: Whither water management? Science 2008, 319, 573–574.
6. Milly, P.C.D.; Dunne, K.A.; Vecchia, A.V. Global pattern of trends in streamflow and water availability in a changing climate. Nature 2005, 438, 347–350.
7. Bates, B.; Kundzewicz, Z.; Wu, S.; Palutikof, J. Climate Change and Water; Intergovernmental Panel on Climate Change: Geneva, Belgium, 2008.
8. Arnell, N.W.; Hudson, D.A.; Jones, R.G. Climate change scenarios from a regional climate model: Estimating change in runoff in southern Africa. J. Geophys. Res. 2003, 108, doi:10.1029/2002JD002782.
9. IPCC. Climate Change 2007: The Physical Science Basis. Contribution of Working Group I to the Fourth Assessment Report of the Intergovernmental Panel on Climate Change; IPCC: Geneva, Belgium, 2007.
10. Madani, K. Hydropower licensing and climate change: Insights from cooperative game theory. Adv. Water Resour. 2011, 34, 174–183.
11. Medellin-Azuara, J.; Harou, J.J.; Olivares, M.A.; Madani, K.; Lund; Howitt, R.E.; Tanaka, S.K.; Jenkins, M.W.; Zhu, T. Adaptability and adaptations of California's water supply system to dry climate warming. Clim. Chang. 2008, 87, 75–90.
12. Madani, K.; Lund, J.R. Estimated impacts of climate warming on California's high-elevation hydropower. Climat. Chang. 2010, 102, 521–538.
13. Koch, F.; Prasch, M.; Bach, H.; Mauser, W.; Appel, F.; Weber, M. How will hydroelectric power generation develop under climate change scenarios? A case study in the upper danube basin. Energies 2011, 4, 1508–1541.
14. Ligare, S.T.; Viers, J.H.; Null, S.E.; Rheinheimer, D.E.; Mount, J.F. Non-uniform changes to whitewater recreation in California's Sierra Nevada from regional climate warming. River Res. Appl. 2011, doi:10.1002/rra.1522.
15. Madani, K.; Lund, J. Modeling California's high-elevation hydropower systems in energy units. Water Resour. Res. 2009, 45, doi:10.1029/2008WR007206.
16. U.S. Energy Information Administration (EIA). International Energy Outlook, 2011. Available online: http://www.eia.gov/forecasts/ieo/ (accessed on 14 February 2012).
17. United Nations Environment Programme (UNEP). Environment Data Explorer, GEO Data Portal, 2010. Available online: http://geodata.grid.unep.ch/ (accessed on 14 February 2012).
18. Food and Agriculture Organization (FAO). Water: Natural Resources Management and Environment Department, 2010. Available online: http://www.fao.org/nr/water/ (accessed on 14 February 2012).
19. Markoff, M.S.; Cullen, A.C. Impact of climate change on Pacific Northwest hydropower. Clim. Chang. 2008, 87, 451–469.
20. Christensen, N.S.; Wood, A.W.; Voisin, N.; Lettenmaier, D.P.; Palmer, R.N. The effects of climate change on the hydrology and water resources of the Colorado River Basin. Clim. Chang. 2004, 62, 337–363.

21. Shongwe, M.E.; van Oldenborgh, G.J.; van den Hurk, B.J.J.M.; de Boer, B.; Coelho, C.A.S.; van Aalst, M.K. Projected changes in mean and extreme precipitation in Africa under global warming. Part I: Southern Africa. J. Clim. 2009, 22, 3819–3837.
22. Filion, Y. Climate change: Implications for Canadian water resources and hydropower generation. Can. Water Rescour. J. 2000, 25, 255–269.
23. Dibike, Y.B.; Coulibaly, P. Hydrologic impact of climate change in the Saguenay watershed: Comparison of downscaling methods and hydrologic models. J. Hydrol. 2005, 307, 145–163.
24. Cherkauer, K.A.; Sinha, T. Hydrologic impacts of projected future climate change in the Lake Michigan region. J. Great Lakes Res. 2010, 36, 33–50.
25. Bergstrom, S.; Carlsson, B.; Gardelin, M.; Lindstrom, G.; Pettersson, A.; Rummukainen, M. Climate change impacts on runoff in Sweden; assessments by global climate models, dynamical downscaling and hydrological modelling. Clim. Res. 2001, 16, 101–112.
26. Meili, Z.; Qian, Y.; Zhihui, L. Climate impacts on hydro-power development in China. Proc. SPIE 2005, 5884, doi:10.1117/12.620728.

There is one supplemental file that is not available in this version of the article. To view this additional information, please use the citation on the first page of this chapter.

CHAPTER 6

Water Constraints on European Power Supply Under Climate Change: Impacts on Electricity Prices

MICHELLE T. H. VAN VLIET, STEFAN VÖGELE, AND DIRK RÜBBELKE

6.1 INTRODUCTION

Hydropower and thermoelectric (nuclear and fossil-fueled) power plants currently contribute to more than 91% of total electricity production in Europe. At present, 74% of total electricity supply is generated by thermoelectric power plants and 17% by hydropower plants (EIA, accessed 2013 for year 2010). The European energy sector therefore strongly depends on the availability of water resources for hydropower generation, and also on the temperatures of water for cooling of thermoelectric power plants. In particular coal-fired and nuclear power plants rely on large volumes of water for cooling. The thermoelectric power sector is compared to other sectors (e.g. agriculture, industry, domestic use) the largest water user in Europe, accounting for about 43% of total surface water withdrawal (EUREAU 2009).

Recent warm, dry summers (e.g. 2003, 2006 and 2009) showed the vulnerability of the European power sector to reduced water availability and high river temperatures. Several thermoelectric power plants were forced to reduce production, because of environmental restrictions on cooling water use, as water availability was low and legal temperature limits were exceeded (Förster and Lilliestam 2010). In particular nuclear power plants were vulnerable to cooling water shortages during recent warm, dry summers. Figure S1 (supplementary information available at stacks.iop.org/ERL/8/035010/mmedia) shows that both nuclear power plants with once-through and recirculation (tower) cooling systems were affected. Hydropower generation was also reduced during prolonged droughts with low water levels, e.g. the droughts of 2002 and 2003 in Scandinavia (Kuusisto 2004). This had distinct economic impacts like increased electricity prices in northern European countries with high hydropower consumption (Kuusisto 2004, Schmidt-Thomé and Kallio 2006). Limited supply of thermoelectric power and increased production costs also increased electricity prices in other parts of Europe during dry, warm summer periods (Boogert and Dupont 2005, McDermott and Nilsen 2011).

Due to climate change, periods with low summer river flows in combination with high water temperatures are expected to occur more frequently in Europe (Van Vliet et al 2013). This is likely to increase environmental restrictions on cooling water use with substantial reductions in power plant capacities for the next 20–50 years in case adaptation measures are not taken (Van Vliet et al 2012b).

Other previous studies showed performance losses of thermoelectric power plants under climate change (Hoffmann et al 2013, Linnerud et al 2011). Hydropower generating capacity will also be affected by altered river flow patterns under climate change (EEA 2012, Lehner et al 2005). Hydropower is currently the main renewable energy source contributing to electricity supply in Europe, and its contribution is anticipated to rise significantly in the next decades (e.g. GEA 2012, Punys et al 2008). It is therefore also important to include the future power plant stock in calculations of changes in hydropower and thermoelectric power generating capacity (Rübbelke and Vögele 2013), and future changes in water demands for electricity generation (Davies et al 2013, Flörke et al 2012, 2011, Kyle et al 2013).

Changes in thermoelectric and hydropower capacity can have important economic consequences, like changes in electricity prices (Mideksa and Kallbekken 2010) and altered import and export balances between different countries (Kopytko and Perkins 2011). This has also previously been demonstrated by Rübbelke and Vögele (2011, 2013) who focused mainly on climate change impacts on nuclear power plants using scenarios that assume slight and more serious water scarcity in Europe.

Here, we assess the impacts of climate change on European electricity production and prices with a focus on both thermoelectric (nuclear and fossil-fuelled) power and hydropower, and using hydrological and water temperature modelling results for future climate. We used simulations of daily river flow and water temperature projections that were produced using a physically based hydrological-water temperature modelling framework with climate model data for 2031–2060 (Van Vliet et al 2012b). These projections of river flows and water temperatures were used in a thermoelectric power and hydropower production model to calculate impacts on power generating capacity. In addition, we explored the effectiveness of adaptation strategies regarding replacements in cooling systems and changes in source of fuels of thermoelectric power plants. Based on the changes in potential electricity generation, we modelled the cost-optimal use of power plants for each European country by including future power plants stocks and taking electricity import and export constraints into account.

6.2 MODELS AND SCENARIOS

The methodological framework of this study is summarized in figure 1. We used daily river flow and water temperature simulations for Europe under current and future climate, which were produced with a physically based hydrological-water temperature modelling framework (Van Vliet et al 2012b). This modelling framework consists of the Variable Infiltration Capacity (VIC) model (Liang et al 1994) and one-dimensional stream temperature River Basin Model (RBM) (Yearsley 2009, 2012). The modelling framework was applied on a daily time step and 0.5° × 0.5° spatial resolution. The geographic extent is defined to include the European continent excluding the northern islands and Ural region. The performance of the modelling framework was previously evaluated for Europe using observed daily river flow and water temperature series for river basins with different characteristics (climate zones, human impacts). This showed an overall realistic representation of the observed conditions for 1971–2000 (Van Vliet et al 2012a, 2012b). Daily simulations of river flow and water temperature were produced for the period 1971–2000 (reference) and for 2031–2060 by forcing the hydrological and water temperature modelling framework with biased-corrected general circulation model (GCM) output (Hagemann et al 2011) for both the IPCC SRES A2 (medium–high) and B1 (low) emission scenarios (Nakicenovic et al 2000) (see supplementary section 1 for details, available at stacks.iop.org/ERL/8/035010/mmedia).

In a next step, we quantified the impacts of changes in river flow and water temperature under climate change on thermoelectric and hydropower generating capacity in Europe. Climate change impacts on hydropower capacity were assessed by calculating gross hydropower potentials according to a similar approach as previously tested by Lehner et al (2005). In this approach, gross hydropower potential is directly calculated from gridded datasets of water availability and elevation differences, without requiring additional data of exact location and installed capacities of hydropower plants (see supplementary section 2 for equation and details, available at stacks.iop.org/ERL/8/035010/mmedia). Lehner et al (2005) concluded that this approach (with use of local elevation differences) can be a good indicator for estimating the relative change in actual hydropower potential. We used gridded simulations of daily river flow under both reference (1971–2000) and future (2031–2060) climate for the SRES A2 (medium–high) and B1 (low) emissions scenario to quantify impacts on gross hydropower potential. Country based changes were subsequently estimated by calculating the mean change in gross hydropower potential for all grid cells within each country.

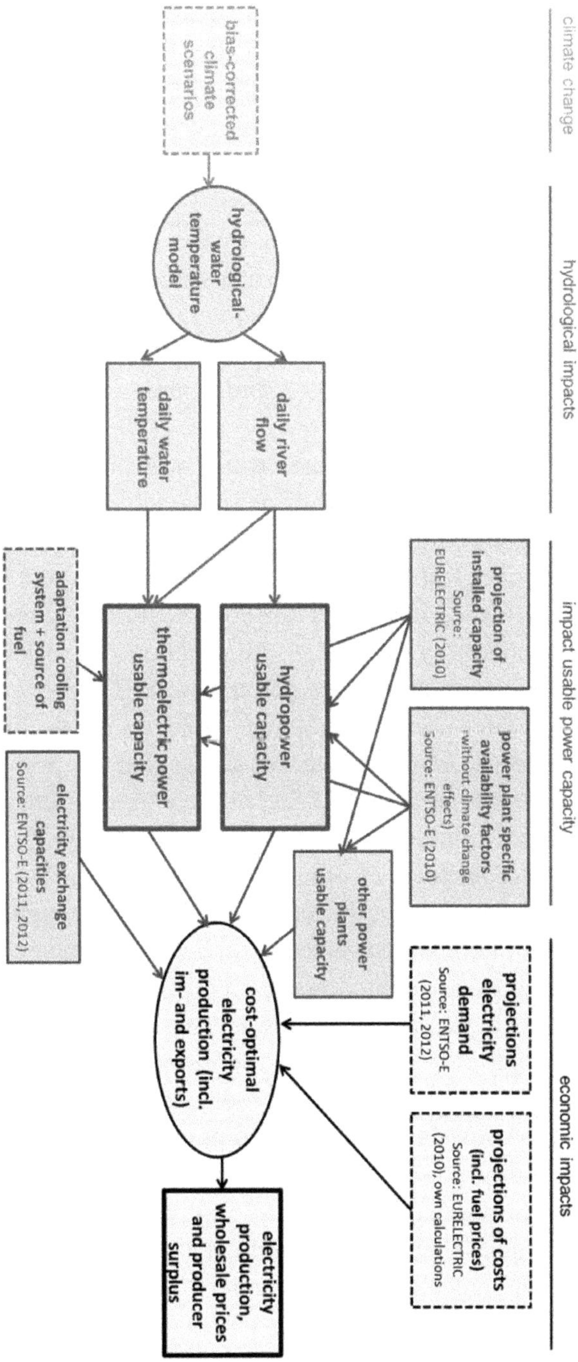

FIGURE 1: Schematic representation of methodological framework

Impacts on thermoelectric power capacity were quantified with a thermoelectric power production model (Koch and Vögele 2009, Rübbelke and Vögele 2011). This model calculates in a first step the water demands of power plants based on their efficiency, installed capacity, cooling system type and the maximum allowed water temperature (increase). In a next step, the useable power plant capacity is calculated based on the estimated daily required water demand, environmental limitations (maximum water temperature (increase) and water withdrawal) and the daily simulations of river flow and water temperature at the power plant site under reference (1971–2000) and future (2031–2060) climate (see supplementary section 3 for details and model equations, available at stacks.iop.org/ERL/8/035010/mmedia).

We focused on 68 thermoelectric power plants in Europe, situated in different European countries and characterized by different sources of fuel and cooling system types (figure 2). For each country we selected thermoelectric power plants with the largest installed capacity and which use river water for cooling. Other criteria for selection were the availability of detailed information of the location, efficiency, cooling system and fuel type, and water temperature limits at the power plant site. Impacts of climate change on daily useable capacity of thermoelectric power plants were assessed for present power plant settings (i.e. current source of fuel and cooling system types), which is denoted henceforth as the "baseline setting" (figure 2). In addition, we also quantified impacts of climate change on useable power capacities considering adaptation in the thermoelectric power sector with regard to replacement of cooling system type and changes in source of fuel. Two additional cases of adaptation were considered; (i) replacement of all once-through by recirculation (tower) cooling systems (denoted henceforth as "adapt cooling"); and (ii) replacement of all once-through by recirculation cooling systems and replacement of all coal-, lignite- and oil-fuelled power plants by gas-fired power plants (denoted henceforth as "adapt cooling + fuel"). Using the relative change in thermoelectric power plant capacities for the different power plant sites in each country we quantified for each country the average impacts of climate change on thermoelectric power under "baseline setting", "adapt cooling" and "adapt cooling + fuel".

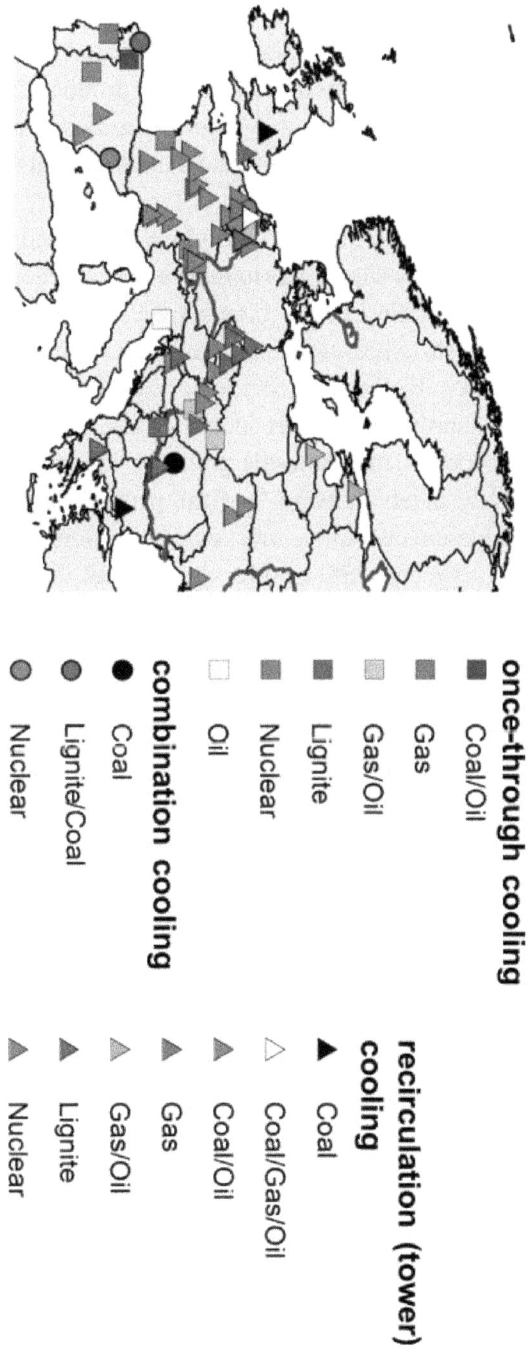

FIGURE 2: Location of power plants, cooling system type (symbols) and source of fuel (colours).

Based on projected changes in thermoelectric and hydropower generating capacity under climate change, we calculated changes in wholesale electricity prices, production and electricity producer surplus. We used scenarios of future installed power plant capacities, electricity exchange capacities, cost figures and electricity demand based on the European Network of Transmission System Operators for Electricity (ENTSO-E 2010, 2011, 2012). Our calculations are based on the assumption that at each point in time electricity supply has to meet electricity demand. In addition, we assumed that existing power plants are used in order of their short-term marginal cost. (Short-term marginal cost comprises operation and maintenance cost as well as fuel cost, ignoring capital costs.) Country specific electricity supply curves were identified taking the cost for electricity imports into account. The wholesale price corresponds to the price where electricity supply meets demand, and the price at the wholesale market equals the production cost of the most expensive power plant in operation. We assumed a price elasticity of 0 for the demand for electricity, because changes in spot market prices have overall small impacts on end-user prices and demands. Equations (1a)–(1c) show the optimization approach (objective function) of our model. Equation (1b) reflects that electricity supply has to meet electricity demand, whereas electricity can be supplied either by using domestic power plants or by importing electricity from a foreign country. Equation (1c) describes the grid capacities, which could be used to export electricity from country n to country m.

$$\min z_t = \sum_n \sum_i c_i \cdot s_{i,n,t} \cdot X_{i,n} + \sum_n \sum_i c_i \cdot imp_{n,m,t} \quad (1a)$$

$$\sum_t s_{i,n,t} \cdot X_{i,n} + \sum_m imp_{n,m,t} \geq d_{n,t} \quad \forall n \quad (1b)$$

$$imp_{n,m,t} \leq NTC_{n,m} \quad \forall (n,m) \quad (1c)$$

where: n,m = index for the country; i = index for power plant type; t = index for time (—); ci = operating costs of power plants of type i (€ MWh−1); $s_{i,n,t}$ = average utilization of power plants of type i in country n, whereas $0 \leq s_{i,n} \leq 1$ (—); $X_{i,n}$ = installed useable capacity of power plants of type i in country n (MW); c_l = costs for transferring electricity from one country to another one (€ MWh^{-1}); $imp_{n,m,t}$ = net-imports of electricity of country n from country m (€ MWh^{-1}); $d_{n,t}$ = electricity demand in country n at time t; $NTC_{n,m}$ = net transfer capacities.

The cost optimization approach used in our dispatch model is implemented in GAMS (general algebraic modelling system) with an hourly time resolution. Prevailing constraints on net transfer capacities ($NTC_{n,m}$) and electricity demand ($d_{n,t}$) are taken into account by using data of ENTSO-E (2011, 2012). The demand ("load") figures of ENTSO-E include electricity demand of private households, industry and other sectors, and transmission losses. In addition, cost figures were used for international trade of electricity reflecting transmissions losses. The model uses data with power plant stock projections for 2030 of EURELECTRIC (2010). Taking the vintage structure of power plants into account the power plants are grouped in 36 different categories. Information of EURELECTRIC was used to draw inferences on power plant efficiencies and calculate power plant specific production cost by taking expected changes in fuel prices into account (EURELECTRIC 2010, IEA 2011a, 2011b). For siting of new power plants, we assumed that companies tend to prefer using existing power plants sites due to advantages regarding the approval procedure for new power plants and infrastructure aspects (e.g. vicinity to consumers, coal mines or gas pipelines, grid connection). Uncertainties regarding electricity demand and power plant availabilities were included by assuming that 5–10% of the generating capacity on country level will be available as spare capacity. Therefore, the available capacity at each point in time will not be used completely, and the mix used for supplying electricity can vary significantly between time slices (e.g. taking into account variations in wind and solar photovoltaics (PV) production).

Our analyses focused on 29 European countries which dominate electricity supply in the Europe. The share of various electricity production techniques for different European countries for the baseline situation is presented in figure S2 of the supplementary information (available at stacks.iop.org/ERL/8/035010/mmedia). The impacts of changes in power plant availability on overall profits were assessed by using a producer surplus approach. Electricity producer surplus reflects the difference between revenues suppliers of electricity obtained from selling electricity and the cost of supplying this electricity. Producer surplus depends not only on the level of the wholesale prices and the production volume but also on the slope of the cost curve (see supplementary section 4, available at stacks.iop.org/ERL/8/035010/mmedia and Rübbelke and Vögele (2013) for details).

6.3 RESULTS

6.3.1 CHANGES IN RIVER FLOW AND WATER TEMPERATURES UNDER CLIMATE CHANGE

Spatial patterns of changes in mean river flow in Europe under future (2031–2060) relative to reference (1971–2000) climate show a strong division between the northern and southern part (figure 3(a)). An increase in mean annual river flow for northern Europe is projected with an average of 3–5% (north of 52 °N), and an average decline for southern Europe of 13–15% (south of 52 °N) for the SRES B1–A2 scenario. Low flow values (10-percentile daily river flow) are projected to decline for almost whole of Europe except for Scandinavia (figure 3(b)). Strongest declines in mean and low flow are mainly projected for southern and south-eastern European countries (Portugal, Spain, Italy, Macedonia, Bulgaria and Greece) with declines of more than 20%.

Increases in mean water temperature are largest (>1 °C) in central Europe (e.g. Switzerland, Austria, Slovenia, Hungary, Slovakia) and south-eastern parts (e.g. Romania, Bulgaria, Croatia, Serbia) (figure 3(c)). The average increase in mean annual water temperatures for the whole European region is 0.6–0.8 °C (SRES B1–A2 for 2031–2060). Overall, larger increases in the high (95th percentile) water temperature range are projected (average of 0.9–1.1 °C). A combination of strong increases in water temperature and declines in low river flow is generally most critical for cooling water use. These conditions are mainly projected for southern, central and south-eastern Europe.

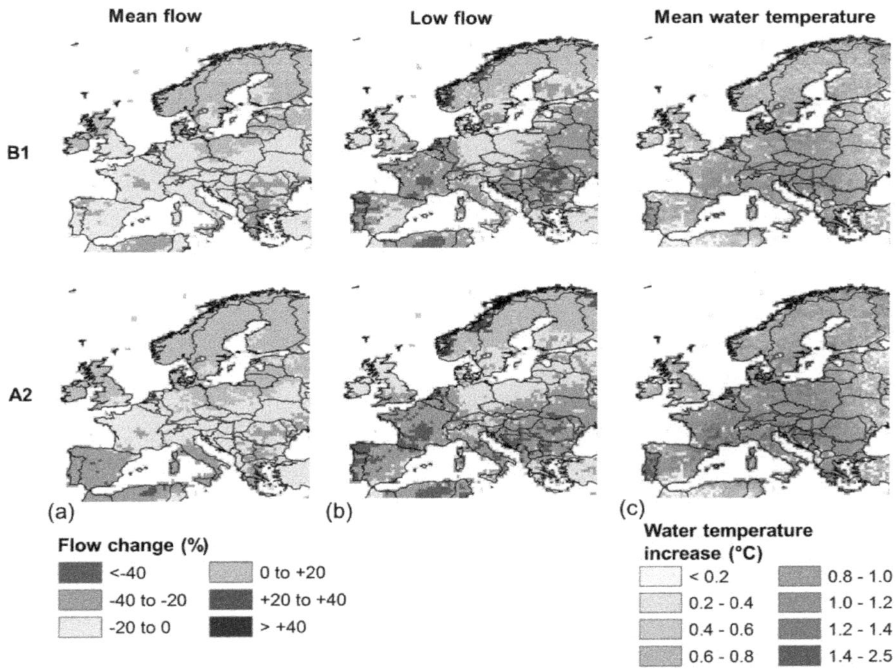

FIGURE 3: Climate change impacts on low river flows and water temperatures in Europe. Projected changes in mean flow (a) and low flows (10th percentile of daily distribution of river flow); (b) and mean water temperatures (c) for the 2031–2060 relative to 1971–2000. Changes are presented using the GCM ensemble mean changes for both the SRES A2 and B1 scenario relative to the reference period.

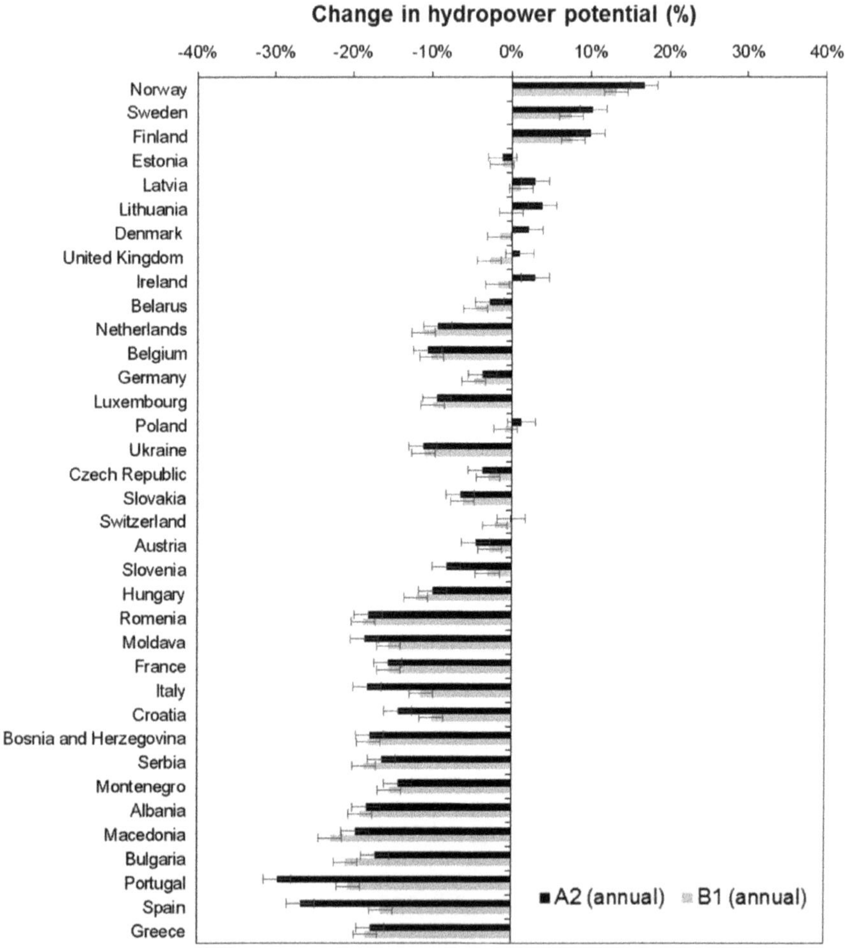

FIGURE 4: Country based statistics of relative (%) changes in mean annual gross hydropower production potential under future climate (2031–2060) relative to current climate (1971–2000). Standard bars represent one standard error.

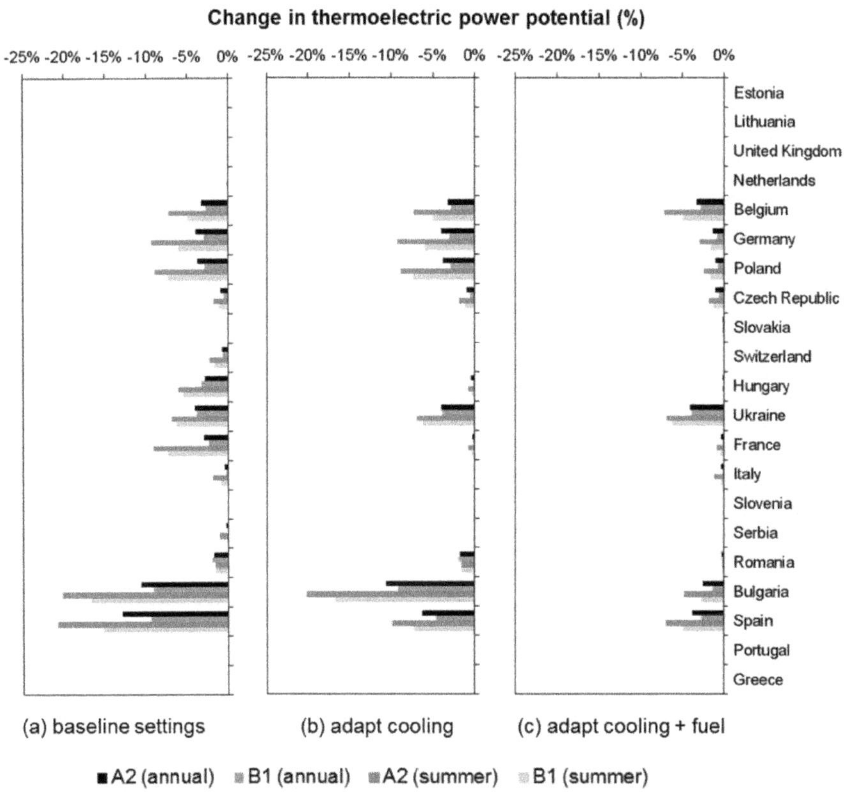

FIGURE 5: Country based statistics of relative (%) changes in thermoelectric power production under future climate (2031–2060) relative to reference climate (1971–2000). Changes are presented on a mean annual basis and for summer period for present power plant setting ('baseline setting') (a), for a scenario of replacement of all once-through by tower cooling systems ('adapt cooling') (b) and replacement of both cooling system and source of fuel ('adapt cooling + fuel') (c).

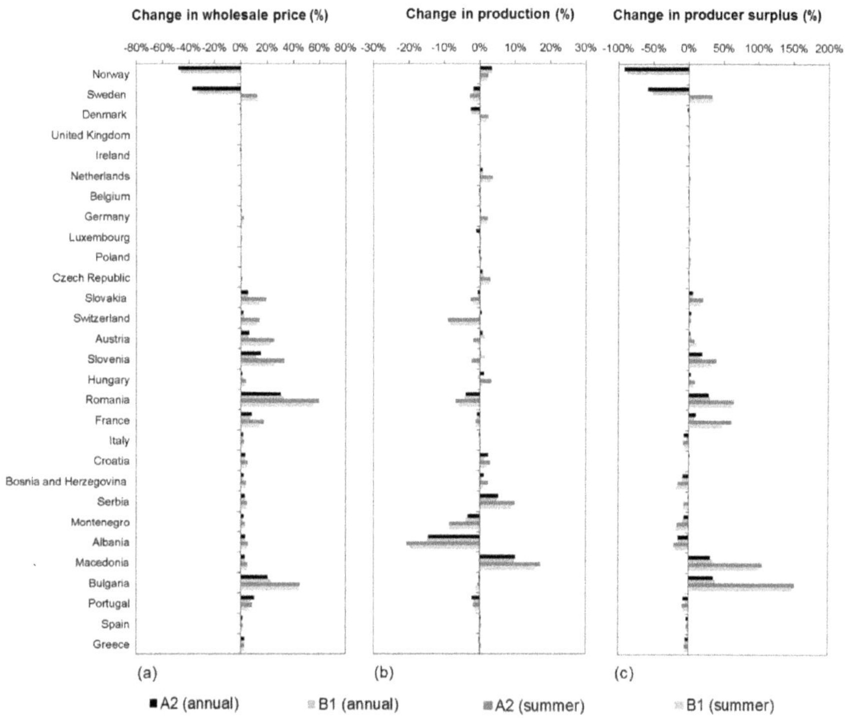

FIGURE 6: Impacts of changes in hydropower and thermoelectric power useable capacity on wholesale prices (a) production (b) and producer surplus (c) on annual mean basis and for summer period.

6.3.2 IMPACTS ON HYDROPOWER AND THERMOELECTRIC POWER GENERATING CAPACITY

Relative change in mean gross hydropower potential largely depicts the projected changes in mean annual river flow. A substantial (>8%) increase in the potential to generate hydropower is projected for northern countries (Norway, Sweden, and Finland) (figure 4). Large (>15%) declines in hydropower potential are projected for southern countries (Portugal, Spain, France) and southeast Europe (Balkan countries like Greece, Bulgaria, Romania, Serbia, Macedonia). There is also a group of countries at mid-northern latitude (50–60 °N) where gross hydropower potential is expected to show moderate changes under future climate (e.g. Latvia, Lithuania, UK and Ireland). In total, the gross hydropower potential of Europe is estimated to decrease on average by 4–5% for the period 2031–2060 (SRES B1–A2) relative to 1971–2000.

Impacts of climate change on thermoelectric power generating capacity were assessed for the present power plant setting ("baseline setting") (figure 5(a)), for a scenario of replacement of all once-through by tower cooling systems ("adapt cooling") (figure 5(b)) and replacement of both cooling system and source of fuel ("adapt cooling + fuel") (figure 5(c)). The largest declines in mean useable capacity under "baseline setting" are estimated for countries in southern and south-eastern Europe. The estimated reduction in summer mean useable capacity is 16–20% for Bulgaria and 15–21% for Spain (SRES B1–A2 for 2031–2060 relative to 1971–2000). Substantial reductions (>5%) in summer mean thermoelectric power capacities are also expected for central European countries (Germany, Poland, Hungary, Ukraine, France, Belgium), where overall high increases in water temperatures combined with strong declines in low summer flow are projected. Replacement of cooling systems and changes in the sources of fuel lead to an overall reduction in the vulnerability of thermoelectric power plants to climate change. For example for power plants in Spain, a replacement of once-through by tower cooling systems ("adapt cooling") decreases the adverse impacts on summer mean useable capacity (reduction of 7–9% for "adapt cooling" compared to 15–21% for "baseline setting" for SRES B1–A2 for 2031–2060). In addition, a replacement of fossil-fuelled power plants by new gas-fired power plants ("adapt cooling + fuel") further lowers reductions in summer mean useable capacity to 5–7%, although impacts are still non-negligible for most countries in southern and central Europe. For Belgium, Czech Republic and Ukraine we found very limited impacts of replacements of cooling systems and sources of fuel, because most power plants in these countries are nuclear or gas-fired power plants that already used recirculation cooling systems under "baseline setting".

6.3.3 IMPACTS ON WHOLESALE PRICES AND DISTRIBUTIONAL CONSEQUENCES

Our calculations for the electricity system in Europe show that overall higher wholesale prices are expected for most countries (figure 6(a)), because the limitations in water availability and exceeded water temperature limits mainly affect power plants with low production cost (e.g. hydroelectric and nuclear power plants). Strongest increases in mean annual wholesale prices are projected for Slovenia (12–15%), Bulgaria (21–23%) and Romania (31–32% for 2031–2060 relative to 1971–2000 for "baseline setting"). Sweden and Norway are exceptions, because mean water availability is projected to increase in these countries, and consequently, more electricity will be produced there by using "low-cost" hydroelectric power plants, putting costlier power plants out of operation. Wholesale prices during summer period are, however, expected to increase for Sweden (with 12%). For most countries, the increases in wholesale electricity prices are higher for summer period than on mean annual basis, because restrictions in cooling water use for thermoelectric power and reductions in water availability for hydropower potential are highest during this season. Strongest increases in wholesale prices during summer are projected for Slovakia (14–19%), France (14–17%), Austria (22–25%), Slovenia (25–33%), Romania (55–60%), and Bulgaria (44–45% for SRES B1–A2 for 2031–2060 relative to 1971–2000).

Overall, the differences between the scenarios (i.e. B1 and A2 emission scenarios, and "baseline settings" and "adapt cooling") are small (see figure 7 for selection of countries). The replacement of coal-fuelled power plants by gas-fuelled power plants is expected to result in high increases in wholesale prices, because the fuel cost of gas-fired power plants are significantly higher than of coal-fired power plants.

The changes in power plant availability and wholesale prices will also affect electricity import and export balances between countries. For instance, countries like Norway will export more electricity, while countries which have no or only expensive excess capacities (e.g. Albania) will import more electricity. Overall, electricity production will be extended in countries like Serbia (mean annual increase of +5%), Macedonia (+10%) and Norway (+4%), while considerable declines in electricity production (up to 15%) are projected for other countries (e.g. Albania, Romania, Montenegro, Sweden) (figure 6(b)).

To assess the overall economic effect for producers, besides changes in production and wholesale prices, changes in the slope of the country specific cost curves were taken into account (Rübbelke and Vögele 2013). Producer surplus will increase considerably in France, Slovenia, Macedonia, Bulgaria and Romania, where increases are more than 10% on mean annual basis and more than 30% for summer period. Strong declines in mean annual producer surplus are expected for Norway and Sweden, which are mainly caused by decreasing wholesale prices of electricity (figure 6(c)).

6.4 DISCUSSION AND CONCLUSIONS

This study shows that the combination of increased water temperatures and reduced summer river flow under climate change is likely to affect both hydropower and thermoelectric power generating capacity in Europe, with distinct impacts on electricity prices. An overall increase in mean hydropower generating potential is expected for northern European countries, but strong declines (>15%) are expected in particular for the southern and south-eastern parts. For whole of Europe, gross hydropower potential is expected to decrease by 4–5% for the period 2031–2060 (SRES B1–A2) relative to 1971–2000, which is in line with the 6–12% reduction

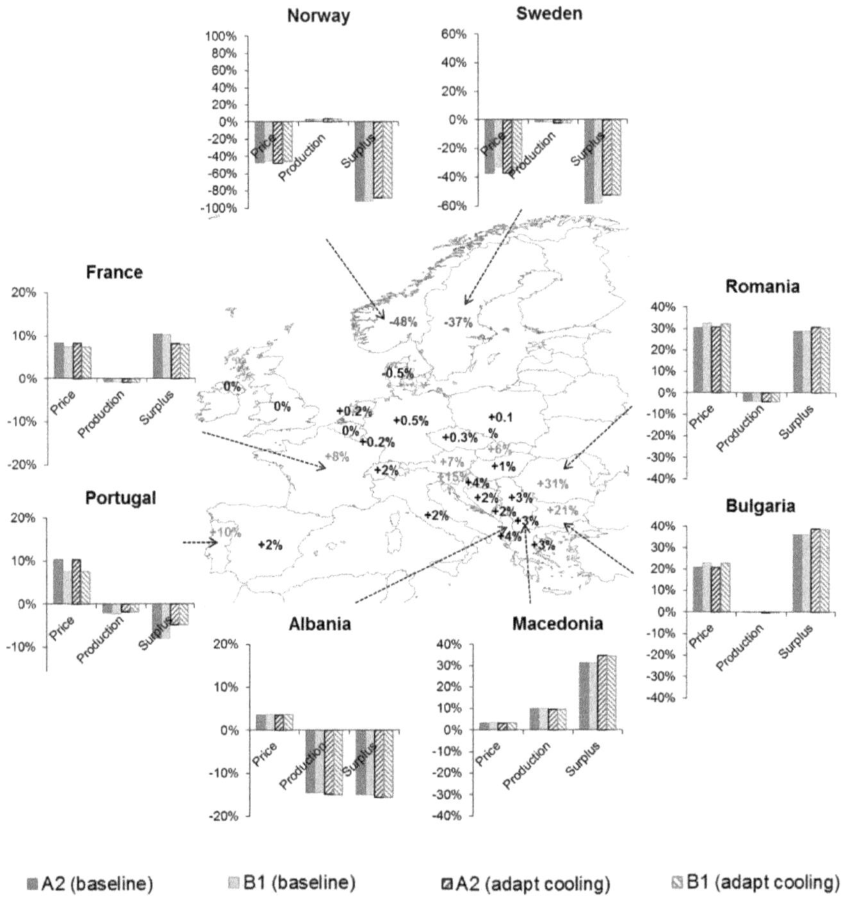

FIGURE 7: Impacts of changes in hydropower and thermoelectric power useable capacity on mean annual wholesale prices, production, and producer surplus for a selection of countries with substantial economic impacts (>10% change). Impacts are presented for both the SRES A2 and B1 emission scenarios and for both the "baseline setting" and "adapt cooling". Values in the maps show the change in mean annual wholesale electricity prices for SRES A2 with "baseline setting" of power plants.

in gross hydropower potential under climate change estimated by Lehner et al (2005) for the period around 2070.

Useable capacity of thermoelectric power is expected to be most strongly impacted in countries in the central, southern and south-eastern part of Europe, where strong declines in low summer river flow are projected in combination with large increases in water temperature. This is expected to increase environmental restrictions on cooling water uses, and can result in substantial reductions in power plant capacities during summer (up to 16–20% for Bulgaria and 15–21% for Spain for SRES B1–A2 for 2031–2060). Replacement in cooling systems and changes in sources of fuel (increased efficiency) reduce water demand of thermoelectric power plants for cooling, and therefore decrease the vulnerability to climate change and aggravated cooling water shortage. However, estimated reductions in power plant useable capacities for power plants with recirculation (tower) cooling systems were smaller but still non-negligible, especially for countries in southern, central and south-eastern Europe.

The cost-optimal use of power plants was calculated on a country level. In our approach we assumed that a liberalized competitive electricity market prevails, but it should be noted that some electricity markets in European countries are still highly regulated and marginal cost pricing is only partly observed. In addition, it should be stressed that uncertainty prevails concerning relative pricing depending on prices of fossil fuels and environmental regulations in the future.

Overall, higher wholesale prices are projected for most European countries, as limitations in water availability mainly affect power plants with low production costs there. Exceptions are Sweden and Norway where water availability for hydropower generation will increase. The impacts of changes in power plant availabilities on national electricity supply systems also depend on the future structure of the power plant stocks. As long as 'inexpensive' excess capacities are available, decreases in power plant availabilities will be bearable, but significant changes in wholesale prices are projected in case no (or only expensive) excess capacities can be used. Although total welfare of market actors declines, some individual market actors might benefit from the impacts of more warm, dry summers on electricity generation patterns (e.g. in France, Slovenia, Macedonia, Bulgaria and Romania). These beneficiaries are not on the consumer side, but on the supply side.

Overall, more electricity will be traded with changes in power plant availabilities in Europe under future climate and changes in power plant stock. Autonomous adaptation via the European electricity market provides opportunities to partly compensate for the loss of power generating capacity in one subsector or plant location by an increase in power generation in another sector or location (Rübbelke and Vögele 2013). However, considering the high shares of hydropower, coal-fuelled and nuclear-fuelled power plants in most European countries, the vulnerability to declines in summer river flow and increased water temperatures can be high. Planned adaptation strategies are therefore highly recommended, especially in the southern, central and south-eastern parts of Europe, where overall largest impacts on thermoelectric and hydropower generating capacity are projected under climate change. Considering the high investments costs (EPRI 2011), retrofitting or replacement of power plants might not be beneficial from the perspective of individual power plant operators, although the social benefits of adaptation could be substantial.

An increased diversification in the electricity sector, with a larger contribution of renewable energy resources that are independent from water availability and water temperature (e.g. solar PV, wind power), might reduce the vulnerability of the electricity sector to climate change, although wind power can also be negatively affected by climate change (Breslow and Sailor 2002). Overall, solar PV and wind power have low production costs, but require considerable backup capacities to compensate for fluctuations in the electricity generation potential. In addition, grid extensions might be required for ensuring electricity supply if the shares of these electricity production technologies increase significantly. Considering the long design life of power plant infrastructure, short-term adaptation strategies are highly recommended to prevent misallocation of resources and to assure future energy security.

REFERENCES

1. Boogert A and Dupont D 2005 The nature of supply side effects on electricity prices: the impact of water temperature Econ. Lett. 88 121–5
2. Breslow P B and Sailor D J 2002 Vulnerability of wind power resources to climate change in the continental United States Renew. Energy 27 585–98
3. Davies E G R, Kyle P and Edmonds J A 2013 An integrated assessment of global and regional water demands for electricity generation to 2095 Adv. Water Resour. 52 296–313
4. EEA 2012 Climate Change, Impacts and Vulnerability in Europe 2012: An Indicator-Based Report (Copenhagen: European Environment Agency)
5. ENTSO-E 2010 Data Portal: Country Data Packages (www.entsoe.eu/data/data-portal/country-packages/)
6. ENTSO-E 2011 NTC Values (www.entsoe.eu/resources/ntc-values/ntc-matrix)
7. ENTSO-E 2012 Ten-Year Network Development Plan 2012 Brussels—Project for Consultation (available at www.entsoe.eu/)
8. EPRI 2011 National cost estimate for retrofit of US power plants with closed-cycle cooling Technical Brief—Clean Water Act Fish Protection Issues (Palo Alto, CA: Electric Power Research Institute) pp 1–2 (available at www.epri.com/)
9. EUREAU 2009 EUREAU Statistics Overview on Water and Wastewater in Europe 2008 (Brussels: European Federation of National Associations of Water and Wastewater Services)
10. EURELECTRIC 2010 Power Statistics (Brussels: Union of the Electricity Industry—EURELECTRIC)
11. Flörke M, Barlund I and Kynast E 2012 Will climate change affect the electricity production sector? a European study J. Water Clim. Change 3 44–54
12. Flörke M, Teichert E and Bärlund I 2011 Future changes of freshwater needs in European power plants Manag. Environ. Qual.: An Int. J. 22 89–104
13. Förster H and Lilliestam J 2010 Modeling thermoelectric power generation in view of climate change Reg. Environ. Change 10 327–38
14. GEA 2012 Global Energy Assessment—Toward a Sustainable Future (Cambridge: Cambridge University Press) (Laxenburg: International Institute for Applied Systems Analysis)
15. Hagemann S, Chen C, Haerter J O, Heinke J, Gerten D and Piani C 2011 Impact of a statistical bias correction on the projected hydrological changes obtained from three GCMs and two hydrology models J. Hydrometeorol. 12 556–78
16. Hoffmann B, Häfele S and Karl U 2013 Analysis of performance losses of thermal power plants in Germany—a system dynamics model approach using data from regional climate modelling Energy 49 193–203
17. IEA 2011a World Energy Outlook 2011 (Paris: OECD/IEA)
18. IEA 2011b World Energy Outlook—Investment Costs (www.worldenergyoutlook.org/investments.asp)
19. Koch H and Vögele S 2009 Dynamic modelling of water demand, water availability and adaptation strategies for power plants to global change Ecol. Econ. 68 2031–9

20. Kopytko N and Perkins J 2011 Climate change, nuclear power, and the adaptation-mitigation dilemma Energy Policy 39 318–33
21. Kuusisto E 2004 Droughts in Finland—past, present and future Hydrol. Days 2004 143–7
22. Kyle P, Davies E G R, Dooley J J, Smith S J, Clarke L E, Edmonds J A and Hejazi M 2013 Influence of climate change mitigation technology on global demands of water for electricity generation Int. J. Greenhouse Gas Control 13 112–23
23. Lehner B, Czisch G and Vassolo S 2005 The impact of global change on the hydropower potential of Europe: a model-based analysis Energy Policy 33 839–55
24. Liang X, Lettenmaier D P, Wood E F and Burges S J 1994 A simple hydrologically based model of land-surface water and energy fluxes for general-circulation models J. Geophys. Res.—Atmos. 99 14415–28
25. Linnerud K, Mideksa T K and Eskeland G S 2011 The impact of climate change on nuclear power supply Energy J. 32 149–68
26. McDermott G R and Nilsen Ø A 2011 Electricity prices, river temperatures and cooling water scarcity Discussion Paper Series in Economics 18/2011 (Bergen: Department of Economics, Norwegian School of Economics)
27. Mideksa T K and Kallbekken S 2010 The impact of climate change on the electricity market: a review Energy Policy 38 3579–85
28. Nakicenovic N et al 2000 Emissions Scenarios. A Special Report of Working Group III of the Intergovernmental Panel on Climate Change (Cambridge: Cambridge University Press)
29. Punys P, Tirunas D, Bagdziunaite-Litvinaitiene L and Pelikan B 2008 The impact of climate change on Kaunas hydropower plant power production 7th Int. Conf. 'Environmental Engineering' pp 674–9
30. Rübbelke D and Vögele S 2011 Impacts of climate change on European critical infrastructures: the case of the power sector Environ. Sci. Policy 14 53–63
31. Rübbelke D and Vögele S 2013 Short-term distributional consequences of climate change impacts on the power sector: who gains and who loses? Clim. Change 116 191–206
32. Schmidt-Thomé P and Kallio H 2006 Natural and technological hazard maps of Europe Natural and Technological Hazards and Risks Affecting the Spatial Development of European Regions (Geological Survey of Finland, Special Paper 42) pp 17–63
33. Van Vliet M T H, Franssen W H P, Yearsley J R, Ludwig F, Haddeland I, Lettenmaier D P and Kabat P 2013 Global river discharge and water temperature under climate change Glob. Environ. Change 23 450–64
34. Van Vliet M T H, Yearsley J R, Franssen W H P, Ludwig F, Haddeland I, Lettenmaier D P and Kabat P 2012a Coupled daily streamflow and water temperature modelling in large river basins Hydrol. Earth Syst. Sci. 16 4303–21
35. Van Vliet M T H, Yearsley J R, Ludwig F, Vögele S, Lettenmaier D P and Kabat P 2012b Vulnerability of US and European electricity supply to climate change Nature Clim. Change 2 676–81

36. Yearsley J R 2009 A semi-Lagrangian water temperature model for advection-dominated river systems Water Resources Res. 45 W12405
37. Yearsley J R 2012 A grid-based approach for simulating stream temperature Water Resources Res. 48 W03506
38. Export references:
39. BibTeX RIS

CITATIONS

1. Vulnerabilities and opportunities at the nexus of electricity, water and climate Peter C Frumhoff et al Environmental Research Letters 2015 10 080201 IOPscience
2. A review of water use in the U.S. electric power sector: insights from systems-level perspectives Rebecca S Dodder Current Opinion in Chemical Engineering 2014 5 7
3. Climate change impacts on hydropower in the Swiss and Italian Alps Ludovic Gaudard et al Science of The Total Environment 2013

CHAPTER 7

How Will Hydroelectric Power Generation Develop under Climate Change Scenarios? A Case Study in the Upper Danube Basin

FRANZISKA KOCH, MONIKA PRASCH, HEIKE BACH, WOLFRAM MAUSER, FLORIAN APPEL, AND MARKUS WEBER

7.1 INTRODUCTION

Since climate change will certainly increase global air temperature, considerable regional impacts on the availability of water resources will occur concerning quantity and seasonality [1]. This will affect all kinds of water users, water suppliers and water management structures. Hydropower has in total a large proportion of the world's energy production, which comprised 15% of the world's total electric energy generation in 2008 [2] following the fossil energy resources carbon, mineral oil and natural gas. In regions with high precipitation rates and steep elevation gradients, like the European Alps, hydropower represents the main electric energy supply.

Alpine countries such as Austria and Switzerland supply over the half of their electric energy mix with hydropower. In Austria, 62% of the internal gross electricity supply was produced by hydropower for the year 2009 [3] and in Switzerland 56.5% for the year 2011 [4]. Moreover, recent significant worldwide increases in hydropower capacity are projected, whereby, the capacity growth in 2008 was second only to wind power [2]. The future development of hydroelectric power generation and its sensitivity to climate change is a relevant and prevailing issue but is not yet well discussed albeit its importance [5]. Hydropower will be impacted by climate change in a varying degree depending on the region and hydropower type; thereby runoff-river power plants as well as reservoir hydropower plants will be affected. Whereas accumulation hydropower production is rather robust to climate variability due to the possibility of adaptation in storage management strategy, runoff-river power plants are directly affected by climate change impacts on runoff in its quantity and seasonality. Regarding reservoir power plants, production strategies can be modified by taking the advantage of the reservoir storage volume and adapting them to climate change effects and possible future changes in electricity demand. However, adaptation is only possible to a certain degree, because they will also be affected by changes in the mean annual runoff and only have limited elasticity in their management plans due to a seasonal runoff shift.

Hydropower as a renewable and sustainable electric energy provider is closely linked to the hydrological situation of a certain region. Hydroelectric power generation depends largely on the regional catchment-based water balance and reacts sensitively to changes of the hydrological cycle in respect to water quantity and seasonality. Seasonal and quantitative changes in precipitation and evapotranspiration lead to inter-annual changes in runoff and hydrological storage, e.g., soil water content, snow cover and glaciers as well as their mean annual amount. Moreover, hydrological periodicities of low-flows and floods as well as the snow and ice storage, especially in mountainous areas, play an important role for runoff generation. Under climate change conditions diverse changes of the mentioned water balance components are expected for the European Alps in the next decades, which are considered by several recent studies [6–18]. The hydrological changes can be expected to consist of are considerable changes in seasonal precipitation patterns indicating an increase in winter and a decrease in summer precipitation, a general increase of mean annual evapotranspiration, a decrease in mean annual runoff and seasonal changes of runoff regimes as well as a decrease of the snow and ice storage. Regarding runoff regimes, a future decrease in snow storage and therefore also in snow-melt leads to an increasing pluvial character until 2060 for the whole Upper Danube basin [11]. The future start of the snowmelt period will be earlier in the year leading to a shift of the hydrological regime and of the seasonal runoff peak. Only in the glaciated alpine head-watersheds, changes of the ice storage play a considerable role [10]. As analyzed in various mountainous catchments, the latter undergo a transition from an ice-melt dominated runoff regime in the past to a more snow-melt dominated regime in the future with a significant shift of the runoff peak from summer to spring [10,11,13,19,20].

Regarding research on future impacts on hydropower, most existing studies analyse the future development of the water cycle in terms of runoff and project changes to the future hydroelectric power generation. Various studies show a mean annual decline of runoff as well as seasonal runoff changes for different regions of the European Alps thus forecasting a future reduction of the mean annual hydroelectric power generation [5,7,21–24]. Besides for the European Alps, studies on climate change impacts on hydropower were carried out also for other regions of the world. For example, regarding the western U.S., several studies [25–28] came to similar findings in their respective region, showing a future decline in mean annual hydroelectric power generation and a seasonal shift due to changes of the hydrological regime. Regarding the European Alps, Schaefli et al. [5] showed a decline in runoff and subsequent hydroelectric power generation for the time period 2070–2099 in a single highly glaciated catchment of a reservoir power plant in Switzerland which was mainly triggered by a decrease in precipitation, glacier retreat and an increase in evapotranspiration. Further, Stanzel and Nachtnebel [21] estimated the changes of the water balance and attributed hydroelectric power generation for the entire Austrian territory using hydrological model outputs resulting in a decrease of 6 to 15% depending on the climate scenario for the time period 2025 to 2075. Due to climate change impacts on seasonal water availability and runoff regimes, an increase in hydroelectric power generation was observed in the winter, whereas a decrease could be found in the summer. Another recently published Austrian study [7] confirms the shift from summer to winter production, but shows only small mean annual changes until 2050. Nonetheless, most of the studies do not directly model the changes of hydropower under altered climate conditions but rather try to assign it to altered runoff conditions.

In this study, the hydroelectric power generation is calculated directly as model output on a high temporal resolution of one hour for single hydropower plants within a large scale catchment. This means that all hydropower plants of the watershed can be considered in a parallel way with their individual parameterisations and the runoff situation for each model time step. The future development of hydroelectric power generation was analysed in detail for the mountainous Upper Danube basin in Central Europe. The study was carried out within the frame of the interdisciplinary research project GLOWA-Danube [29–31], which explored various physical and social impacts of Global Change on water resources of the Upper Danube basin in a regional focus for the next 50 years. To consider all important hydrological components, which have an impact on the hydroelectric power generation and their development due to climate change, the physically-based and fully spatially distributed model Processes of RadiatiOn, Mass and Energy Transfer (PROMET) [32] was used. It strictly conserves mass and energy and is not calibrated against measured discharge at gauges. PROMET is coupled with a specially developed hydropower module, which calculates the hydroelectric power generation for each hydropower plant above 5 MW bottleneck capacity on an hourly resolution. To cover a possible range of future uncertainties, 16 climate scenarios of four underlying regional climate trends were taken as meteorological drivers for PROMET, which were defined from different ensemble outputs of a stochastic climate generator based on the IPCC-SRES-A1B scenario. The analysis thereafter shows a range of possible future developments of hydroelectric power generation for the entire investigation area and three hydrologically different locations, which diverge in their degree of alpine character.

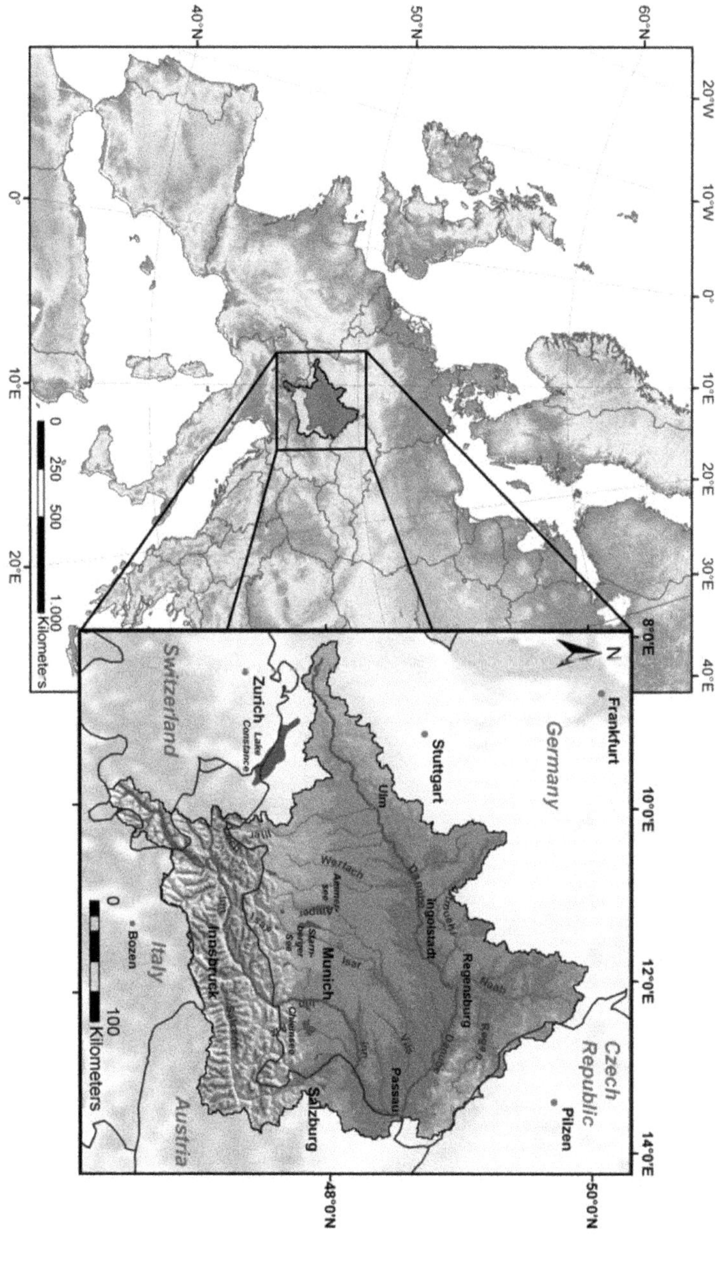

FIGURE 1: Upper Danube basin in Central Europe.

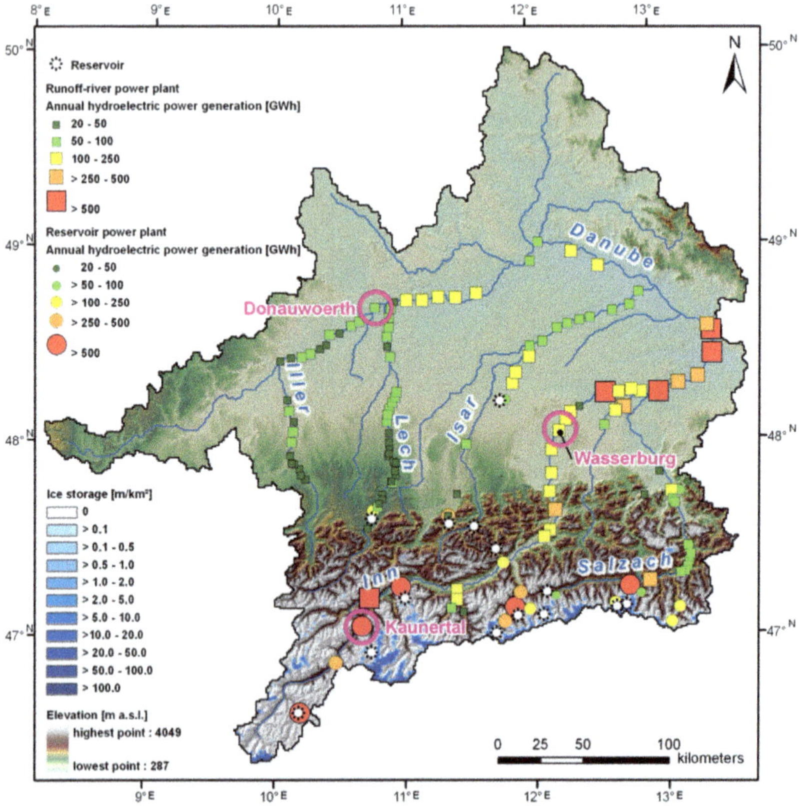

FIGURE 2: The Upper Danube basin with its main rivers, ice storage, hydropower plants and reservoirs. The locations of the hydropower plants Donauwoerth, Wasserburg and Kaunertal, analyzed in Section 4.3, are marked with a pink circle.

7.2 HYDROPOWER IN THE UPPER DANUBE BASIN

The study area, covering parts of Germany and Austria and small parts of Switzerland, Italy and the Czech Republic, is defined by the Achleiten outlet gauge near Passau after the confluence of the rivers Inn and Danube (Figure 1). The Upper Danube is one of Europe's most important catchments and is a main water source for the downstream Danube region until its discharge into the Black Sea.

The catchment is structured by the Alps and the alpine valleys in the southern part, the alpine forelands and the tertiary hills in the middle part, and the Danube lowlands and the mid-altitude mountains of the Bavarian Forest and the Swabian Alb in the northern part. The river Danube is characterized on the one hand by the northern tributaries of the mid-altitude mountains and on the other hand by the southern tributaries originating in the Alps. It is fed to a high portion by the river Inn and its alpine headwaters, which are strongly influenced by snow storage [11,33] and are further characterized by approx. 550 glaciers with an area of 358 km^2 in the year 2000, predominantly situated in the Central Alps [11]. The lowlands of the tributaries Inn, Iller, Lech, Isar and Salzach as well as the downstream Danube heavily depend on significant amounts of water from the alpine head-watersheds and are influenced also by their seasonality. Especially in summer, the Alps play an important role to the lower catchments in runoff delivery [11,12].

The Upper Danube basin covers an area of 76,660 km^2 and is characterized by its mountainous topography with steep elevation gradients. The altitudes range from 4049 m a.s.l at Piz Bernina in the Central Alps to 287 m a.s.l at the Achleiten outlet gauge. Depending on altitude and region, the mean annual temperature varies from −4.7 to +9.0 °C and the mean annual precipitation in the range of 650 mm in the northern part to more than 2000 mm in the Alps. The mean annual evapotranspiration ranges from 100 to 700 mm per year; the mean annual runoff from 150 to 1750 mm per year. Besides these physical gradients, the heterogeneous research area shows strong gradients in social factors and processes. Water resources are intensively and differently used by agriculture, tourism, industry and energy providers [29]. Because of high precipitation and runoff rates as well as steep elevation gradients in the Alps, the Upper Danube basin is ideally suited for hydroelectric power generation. In the Austrian province of Tyrol, hydropower supplies most of the inland electricity demand [34]. In the German Free State of Bavaria which covers more than half of the study area, hydroelectric power generation amounts to about 18% [35]. For this study all hydropower plants in the Upper Danube basin covering a bottleneck capacity of more than 5 MW are considered. This amounts to 140 large power plants, comprising 118 runoff-river power plants and 22 reservoir power plants (Figure 2).

About 70% of the total hydroelectric power generation in the Upper Danube watershed is produced by runoff-river power plants providing base-load energy, whereas about 30% is produced by reservoir power plants presently covering both, mid-load and peak-demand energy, with an elevated electricity demand in winter. In winter the discharge of the river Inn as main Alpine tributary therefore shows a characteristic daily oscillation during working days. The runoff-river power plants are mainly situated at the river Danube and its larger tributaries Iller, Lech, Isar, Inn and Salzach with a range of annual hydroelectric power generation of 20 to more than 500 GWh. Most of the runoff-river power plants generate 50 to 250 GWh. Because of high mean annual runoff, the biggest runoff-river power plants are installed at the river Inn. Nearly all of the reservoir power plants with a range of 50 to 1000 GWh are situated in the southern part in the Central Alps.

7.3 METHODS

To determine the hydroelectric power generation of each hydropower plant, a specific hydropower module was developed. The module is coupled with the hydrological model PROMET [32]. After describing PROMET and the components considering snow- and ice-melt, channel flow and man-made hydraulic structures (Section 3.1), the coupled hydropower module (Section 3.2) is explained in detail and its validation (Section 3.3) is shown. The required meteorological input data for PROMET are explained in Section 3.4.

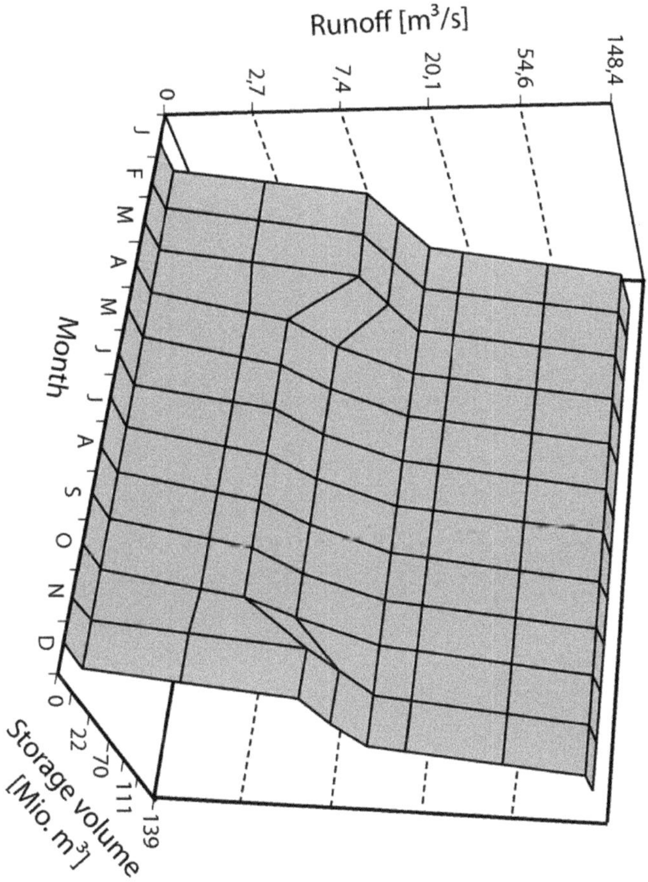

FIGURE 3: Monthly-based look-up table plan of the operation of the Gepatsch reservoir.

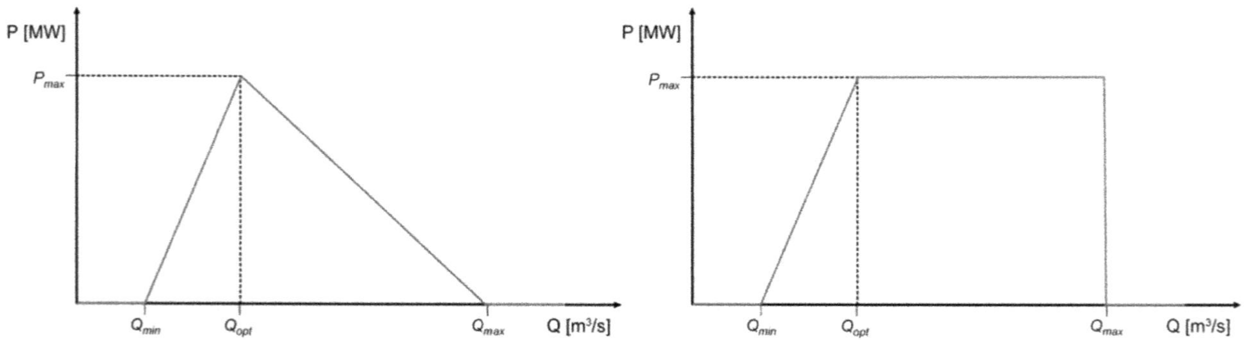

FIGURE 4: Relationship of capacity and runoff for runoff-river power plants (left) and for reservoir power plants (right) used in the model.

7.3.1 THE HYDROLOGICAL MODEL PROMET AND ITS COMPONENTS CONSIDERING SNOW- AND ICE-MELT, CHANNEL FLOW AND MAN-MADE HYDRAULIC STRUCTURES

The hydrological model PROMET is fully spatially distributed, raster-based with a spatial resolution of 1 km^2 and a temporal resolution of one hour in the selected case study. It covers the following components: land surface energy and mass balance, vegetation, snow- and ice-melt, soil hydraulic and temperature, groundwater, channel flow and man-made hydraulic structures. The model is driven by meteorological input data (see Section 3.4). All meteorological, hydrological and land surface components, including land-atmosphere energy and mass exchange, snow and ice accumulation and ablation, flows in the saturated and unsaturated zones, channel flows, and flows through lakes and man-made structures are fully coupled. Thereby PROMET strictly follows the principle of conserving mass and energy fluxes. To simulate these processes, spatial input of topography, land use, glacier extent and ice thickness, soil texture and meteorology are necessary for each grid cell. Moreover, soil, vegetation and runoff parameters, as well as parameters and operation rules for man-made hydraulic structures are implemented in the model. The PROMET input data are described in detail in Mauser and Bach [32]. All model components were validated in detail and the model was and is applied in different watersheds [32,36–42]. PROMET is not calibrated using historical runoff measurements at gauges to preserve its predictive power. Calibration of physically-based hydrological models to historical streamflows generally leads to a good model performance as long as boundary conditions like climate, land use or hydraulic structures stay unchanged. It cannot be expected however that the model performance is equally good under changing future boundary conditions [32,43]. Besides its direct impact on hydrology, climate changes may also change the characteristics of a basin (e.g., through the removal of glaciers) in a way that calibration to past streamflow data, which included the influence of the glaciers, may force the model into inadequate simulations because it is calibrated to a different watershed. Past streamflow data also includes the influences of the existing man-made reservoirs or water transfers. Calibration therefore may become invalid in a strict sense as soon as new structures alter the hydrologic behavior of the watershed. Uncalibrated models offer the possibility to study the effect of adaptation strategies to climate change through, e.g., the installation of new storages or strategic changes in land use (irrigation, deforestation). Nevertheless, the complexity of non-calibrated models should not result in not understanding the underlying processes anymore [39,44]. However, all hydrological impact studies have to deal with uncertainty of, e.g., climate change, the impacts of economic change, population development and different management practices like changes in irrigation and water supply. They therefore rely on ensembles of climate change trajectories and scenarios for the additional future developments, which have to be documented thoroughly.

For this study the model components considering snow- and ice-melt, channel flow and man-made hydraulic structures are most important and will be described shortly in the following. The component considering snow- and ice-melt includes the SUbscale Regional Glacier Extension Simulator (SURGES) module [11,39], which calculates the energy and mass balance and accordingly water equivalent and melt rate of the snow and ice storage. The modelling of snow-melt takes the liquid water storage of a snow pack and rainfall on the surface of the snow cover into account. The module differentiates between solid or liquid state of precipitation using an empirically derived wet-bulb temperature threshold. Melting conditions are indicated by the surface energy balance [33,40]. The details of the glacier topography are parameterized by an area-elevation distribution related to frameworks grid. The three main processes are accumulation, ablation and ice flow. Because changes in the glacier geometry are considered, the module can be used for future long-term simulations [11].

Concerning the channel flow component, it is assumed that each grid cell is part of a channel network, whereby all grid cells are hydraulically connected through topography by using a digital elevation model. Each grid cell then transfers the channel flow to its hydraulic neighbour. Flow velocities and changes of water storage are considered by the Maskingum-Cunge method [45] modified by Todini [46]. The routing component also considers runoff retention in lakes [32].

The man-made hydraulic structures component represents the hydraulic behaviour of reservoirs, which can store channel flow and water transfers. The operation of water transfers as artificial hydraulic connections works with monthly-based operation plans. Its outflow is also operated using a monthly look-up table plan, which translates the storage volume into discharge. This monthly storage-discharge relation allows for a shift of the reservoir inflow and outflow during the course of the year with the main present purpose to store the annual snow- and ice-melt in summer and use it for low-flow augmentation in winter in alpine areas. Each reservoir operates individually by allowing flexible water management strategies, e.g., for hydroelectric power generation or flood protection. The management rules were implemented following the general operation rule suggestions presented by Ostrowski and Lohr [47], which consider normal reservoir operation as well as operation during high and low water availability, taking minimal and maximal discharge capacities, and storage volumes into account. When maximum storage volume is reached, the reservoir switches into a spillway discharge mode.

Each reservoir is characterized by its individual storage volume, storage zoning and mean in- and outflow characteristics. Since detailed information on the actual operation of the reservoirs is not publicly available, all available information from literature, personal contacts and data on reservoir in- and outflows and lake levels was used to set up the individual monthly-based operation scenarios for all reservoirs. The scenarios assume that to use the reservoir storage efficiently, the reservoir mean annual outflow is oriented towards the mean annual inflow, whereby the reservoir filling can vary seasonally between 20 and 80%. Because forecasting is not possible with this module, fillings over 80% are held free due to flood events. Below 20% the outflow is reduced to the set minimal outflow. The operation scenarios refer to present conditions. Throughout the following study of the impact of climate change on hydropower production these reservoir management scenarios were assumed unchanged for the future as a first order approach. A detailed analysis of the adaptation potential of changes in the operation scenarios is beyond the scope of this study. Figure 3 shows the operation scenario for the Gepatsch reservoir. It is the second-largest man-made reservoir in the watershed, located in a glaciated, highly alpine characterized head-watershed and contains a total volume of 139 million m^3 of water. The operation scenario considers the runoff components snow- and ice-melt. This means that in times with high melting rates (May to October) more water will be stored by discharging less runoff at the same storage volume than in months with smaller or zero melting rates (November to April). From Figure 3 it becomes clear that, although principally possible in the module, the operation scenarios do not reproduce the actual operation of these reservoir power plants, which to a certain degree depend on the actual energy demand.

7.3.2 THE HYDROPOWER MODULE

All currently existing hydropower plants with a bottleneck capacity of more than 5 MW (see Figure 2) are implemented in the hydropower module of PROMET to determine hydroelectric power generation. In general, hydroelectric power generation is based on potential and kinetic energy. Therefore the two most important parameters are runoff and hydraulic head. The capacity of each hydropower plant was calculated with an hourly resolution by the following equation:

$$P = \eta \cdot \rho \cdot Q \cdot g \cdot H \qquad (1)$$

where P is the capacity (for a certain time period) (kW), η is the efficiency factor of a hydropower plant (-), ρ is the density of water (kg m^{-3}), g is the gravitational acceleration (m s^{-2}), Q is the runoff (m^3 s^{-1}) and H is the hydraulic head (m). The resulting capacity for each time step was then aggregated to mean daily, annual or decadal hydroelectric power generation values:

$$E = P \cdot t \qquad (2)$$

where E is the hydroelectric power generation (kWh), P is the capacity (for a certain time period) (kW) and t is the time (h).

Each hydropower plant is located within the investigation area at a defined grid cell of 1 km^2. Hence, for each hydropower plant the runoff Q referred to the channel flow component is known for each time step.

It is the only variable component of Equation (1) changing at each time step. Furthermore, for each of the 118 runoff-river power plants and the 22 reservoir power plants (cf. Figure 2), the following parameters were investigated and implemented individually: mean annual hydroelectric power generation, hydraulic head, efficiency factor, maximum capacity and starting year of operation of the power station. All data are based on the parameterization derived from the present.

The left side of Figure 4 illustrates schematically the relationship of capacity P and runoff Q for a runoff-river power plant. In general, an increase in channel flow leads to an increase in capacity until the maximum capacity P_{max} at the optimal runoff Q_{opt} value is reached. For most hydropower plants this point is similar to the maximum discharge of the turbines.

The energy generation starts at a minimum channel flow Q_{min}, which mainly is defined by low-flows and residual flow restrictions. During these situations no energy is produced because low-flows are often not discharged through the turbines. For the runoff-river power plants it is assumed that after achieving P_{max}, the capacity decreases because more runoff leads to a rising downstream water level, which results in turn in a reduction of the hydraulic head. After reaching a set maximum channel flow Q_{max}, the energy production is shut down taking restrictions of energy generation by extreme flood events into account. Overall, this scheme clearly shows that although the capacity is strongly related to the channel flow, low-flow and flood events also have a large impact on capacity.

The relationship of capacity and runoff for a reservoir power plant is handled quite similarly and is shown schematically on the right of Figure 4. The difference is that after reaching Q_{opt}, P_{max} is held constant for a longer time, because a rising downstream water level has less influence. Furthermore, each reservoir power plant underlies monthly-based operating rules (see Section 3.1).

7.3.3 VALIDATION

As the output of the hydropower module—the capacity or rather the hydroelectric power generation—is highly dependent on channel flow and extreme events like low-flows and floods, the hydrological model PROM-

ET was validated using streamflow records at the outlet gauge Achleiten and several other gauges within the catchment on daily and annual time steps. The respective main validation results will be shortly outlined in the following; a detailed description is handled in Mauser and Bach [32]. The validation period for the runoff generated by PROMET is a 33-year model run covering the hydrological years (November-October) 1971 to 2003 with an hourly time step. This takes into account the standard climate period 1971–2000 with the extension of the extremely warm Central European Summer of 2003. The hourly runoff was generated for each 1 km2 pixel and aggregated to daily and annual values for further analysis. Firstly, the annual water balance was compared to the measured annual runoff at the outlet gauge at Achleiten (see Figure 5) through a linear regression analysis. Since the model should ideally reproduce the measured values the model y = a × x was taken as regression hypothesis. Both the slope and the coefficient of determination R^2 should be as close as possible to a value of 1 to ensure that the model reproduces the full dynamic of the measured data without any model bias [37]. The mean modelled runoff at the Achleiten outlet gauge (598 mm/a) compares well with the measured runoff (579 mm/a) [32]. Secondly, the validation of the short term runoff dynamics the measured versus modelled average daily runoff of the 33-years period was also compared at selected gauges in the watershed, by calculating slopes of linear regression lines forced through the origin, coefficients of determination and Nash-Sutcliffe efficiency coefficients [48]. The slopes of the regression lines detect systematic biases in the representation of the natural runoff dynamic by the model. The coefficients of determination give an indication of the amount of variance of the measured data, which is captured by the model simulation. The Nash-Sutcliffe efficiency coefficient compares the mean square error generated by a particular model simulation to the variance of the target output sequence. It is defined as:

$$E = 1 - \frac{\sum_{t=1}^{T}(Q_0^t - Q_m^t)^2}{\sum_{t=1}^{T}(Q_0^t - \overline{Q_0})^2} \quad (3)$$

where E is the Nash-Sutcliffe efficiency coefficient, Q_0, the measured runoff (m^3/s), Q_o the mean measured runoff (m^3/s), the modelled runoff Q_m (m^3/s), T the time period and t the selected time step. An efficiency of 1 corresponds to a perfect match of modelled discharge to the observed data. An efficiency of 0 indicates that the model predictions are as accurate as the mean of the observed data, whereas efficiency less than zero occur when the observed mean is a better predictor than the model. The closer the model efficiency is to 1, the more accurately the model reproduces the data. This normalized measure is used to assess the predictive power of hydrological models; however it does not measure how good a model is in absolute terms. Additionally it should be mentioned that reliability of the Nash-Sutcliffe efficiency coefficient depends on the seasonality of the time series. The model performance for strongly seasonal time series, e.g., glacial regimes, can be overestimated. For time series with small fluctuations around the mean value, this coefficient is a rather good predictor [49].

This analysis was executed at a broad range of different gauges of sub-watersheds at different places in the Upper Danube basin and is summarized together with the results from the analysis of the annual runoffs in Table 1. In general the slopes of the linear regression are close to 1 and the coefficients of determination are high, which indicate that the inter-annual and daily runoff dynamics are well captured by PROMET, both, at the outlet in Achleiten as well as in the sub-watersheds. The coefficient of determination is slightly higher for larger (sub-)watersheds. Accordingly, for the smallest selected watershed the 1 km × 1 km resolution may not be sufficient for this model's settings and input data. Extreme events, namely low-flows and floods, have influence on the hydroelectric power generation of the hydropower module. The validation of low-flows and floods was carried out at the Achleiten outlet gauge for the same time period. The flood analysis considered the annual peak discharge, whereas the low-flow analysis considered the annual lowest 7-day average flow. Regarding the flood analysis a fairly stable relation was found, nevertheless an overestimation by 16% of the simulated values was shown because of neglecting inundations and dam breaks during very large floods. However, the simulated and measured low-flow frequency correlated very well. This also applies to the return periods of both extreme events [32].

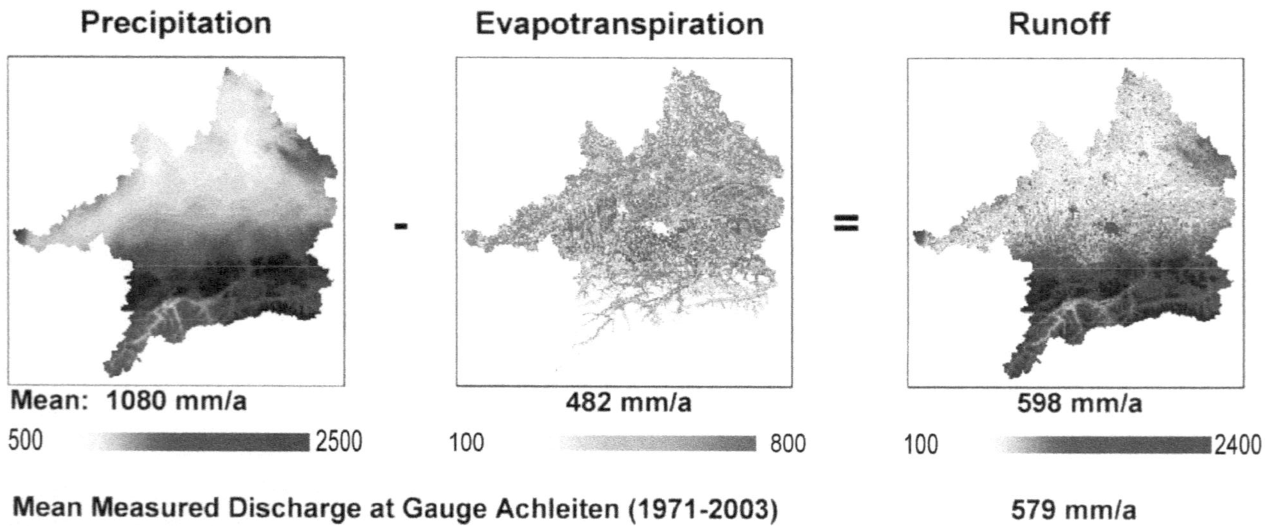

FIGURE 5: Mean modelled water balance in the Upper Danube watershed during the period 1971 to 2003 and mean measured discharge at the Achleiten outlet gauge (based on Mauser and Bach [32]).

TABLE 1: Statistical analysis by calculating slopes of linear regression lines, coefficients of determination and the Nash-Sutcliffe coefficient (only for the daily values) for the linear correlation between the modelled and measured annual and daily runoff of selected (sub-)watersheds in the Upper Danube basin in the period 1971–2003 (based on Mauser and Bach [32]).

			Annual Values		Daily Values		
Gauge name	River	Size of (sub-) watershed	Slope of linear regression	Coefficient of determination	Slope of linear regression	Coefficient of determination	Nash-Sutcliffe efficiency coefficient
Achleiten	Danube	76,660 km^2	1.05	0.93	1.03	0.87	0.84
Hofkirchen	Danube	46,496 km^2	1.12	0.93	1.11	0.87	0.81
Dillingen	Danube	11,350 km^2	1.14	0.93	1.13	0.84	0.72
Oberaudorf	Inn	9715 km^2	0.99	0.80	0.94	0.81	0.80
Plattling	Isar	8435 km^2	1.03	0.88	1.08	0.75	0.47
Laufen	Salzach	6112 km^2	0.93	0.85	0.86	0.85	0.80
Heitzenhofen	Naab	5431 km^2	1.01	0.86	0.99	0.78	0.79
Weilheim	Ammer	607 km^2	1.09	0.88	0.98	0.63	0.69

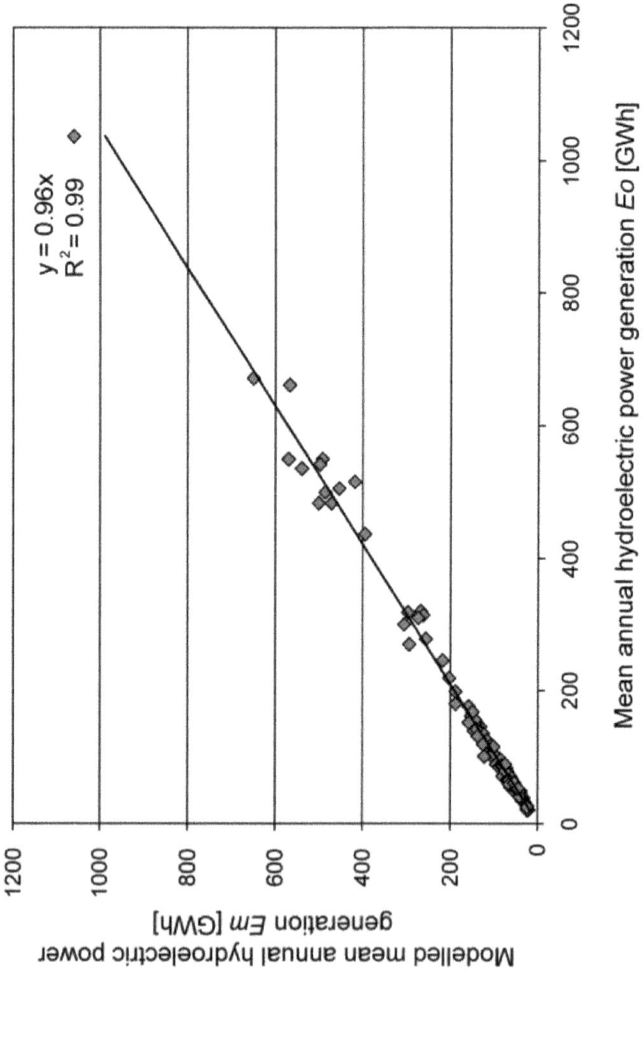

FIGURE 6: Validation of the mean annual hydroelectric power generation by comparing the model output data (E_m) with information from the hydropower plant operators (E_o) for all considered hydropower plants in the Upper Danube basin for the time period 2000–2006.

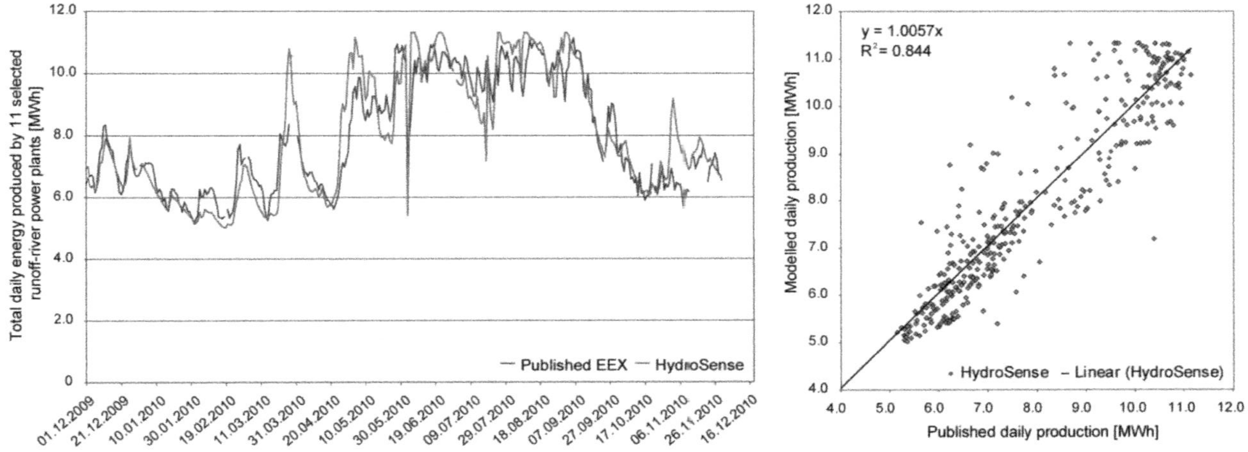

FIGURE 7: Total daily sum of 11 analysed runoff-river power plants located at the rivers Danube and Inn. Comparison of modelled HydroSense and published EEX hydroelectric power generation data.

In general, PROMET, which is not adjusted to observed runoff, is able to model the flow within the channel network and the extreme events low-flows and floods, by agreeing well up, to very well with measured values. Since the runoff is the only variable input of Equation (1) (see Section 3.2), the hydroelectric power generation is well displayed. The mean annual hydroelectric power generation E_m calculated with PROMET has been validated for each power plant with the mean annual hydroelectric power generation E_o published by the hydropower plant operators.

This data was further confirmed by literature, websites and technical papers. The conducted validation of the mean annual values shows a very good correlation with a high coefficient of determination R^2 (0.99) on a long term basis (Figure 6).

The time period 2000 to 2006 was chosen, because since 2000 nearly all hydropower plants integrated in the model had started with the energy production and the meteorological driver data for the model was available until 2006 (see Section 3.4). Regarding the relationship E_m/E_o the mean (0.96) is close to the value 1. The root mean square error (0.08) and the variance (0.01) have low values. This statistical analysis shows that the hydropower plants in the hydropower module are well reproduced for the mean annual values by the simulation.

To further prove the hydroelectric power generation calculated with PROMET, daily production rates as published by the EEX energy stock exchange were consulted and used for validation of the model results. Validation on a finer temporal resolution was performed within the "HydroSense" service by VISTA Remote Sensing in Geosciences GmbH, providing analyses and forecasts of hydropower generation based on PROMET calculation on an operational basis [50]. Based on a dense net of meteorological stations, provided by the private weather data service EWC, hourly and daily results of modeled runoff and hydropower production were created. As validation datasets, hourly published energy production of the EEX energy stock exchange [51] for the year 2010 was taken into consideration. These published datasets are based on a voluntary commitment of the main energy providers in Germany and Austria. Due to publishing structure, only the sum of energy production of 11 river-runoff power plants is provided by the EEX database. A comparison of the cumulated daily energy production for these 11 runoff-river power plants modelled with PROMET at the rivers Danube and Inn for the year 2010 is presented in Figure 7. Modelled daily production (red line) follows the published values (blue line) during the entire year and shows a very good correlation ($y = 1.0057x$; $R^2 = 0.84$).

A further analysis, comprising single power plants or reasonable hourly resolution, will be possible when more detailed information on energy production for river-runoff and reservoir hydropower, will be provided by the energy providers.

7.3.4 METEOROLOGICAL INPUT DATA AND CLIMATE TRENDS

The hydrological model PROMET requires meteorological driver data for each grid cell and each time step. To conduct past and future simulations, data for a past period (1960–2006) and for a future period (2011–2060) were set up during the GLOWA-Danube project. The generation and the specifications of the meteorological input data will be described in the following sections.

The past meteorological data set is derived from 277 climatologic stations from the standard network of the German (DWD) and Austrian (ZAMG) Weather Services. The investigated meteorological variables for each grid cell on a 1 km^2 scale and an hourly time step are precipitation, air temperature, humidity, radiation, horizontal wind speed and air pressure. Therefore, firstly a cubic spline interpolation was used to generate hourly values out of the three standard daily records (7 a.m., 2 p.m. and 9 p.m.) of each meteorological station. Secondly, the spatial interpolation was carried out taking altitudinal gradients and a digital terrain model into account (see [32,36]).

The meteorological data set for the future was generated by using a stochastic climate generator [32,52], recombining the historical dataset, considering statistically different predefined climate trends. This generator is classified as a stochastic nearest neighbour climate generator similar to the approaches of Orlowsky et al. [53], Yates et al. [54], Buishand and Brandsma [55] and Young [56]. It produces a likely realisation of future climate and not synthetic weather data for regions with sparse data available like WGEN [57] or LARS-WG [58,59]. It is assumed that the annual course of a year can be decomposed into weeks represented by temperature means, precipitation sums and their covariance. This climate generator produces weekly sequences of temperature means, precipitation sums and their covariance out of historic data and reassembles them randomly by an underlying climate trend on temperature and precipitation for the future time period. The outputs are new, synthetic time series with the same temporal and spatial resolution as the input data, whereby the physical relations between the meteorological variables are restored. At present stage, future precipitation can be better represented with the climate generator than with most direct regional climate model (RCM) simulations. Another advantage of the climate generator is that the future scenario climate data need no bias correction because it uses the change signal and not RCM data series directly. It should be kept in mind, however, that the general statistical relationship of temperature and precipitation is assumed not to change in future [32]. Regarding the limits of the climate change generator, e.g., a future change in weather patterns, however, can not be considered appropriately with this tool.

To identify possible future changes and to cover a plausible range of uncertainties in regional climate development, 16 climate scenarios, resulting from different ensemble outputs of the stochastic climate generator, were taken as meteorological drivers for the period 2011–2060. All climate scenarios are based on the global IPCC-SRES-A1B emission scenario, which shows a mean development of greenhouse gas emissions due to mid-line economic growth [1], and are part of four regional climate trends. For each of the four trends, four climate scenarios have been selected on statistical criteria. They follow different approaches and are named thereafter IPCC regional, REMO regional MM5 regional, and Extrapolation. The IPCC regional climate trend is based on the results from 21 global climate models for Central Europe presented in the latest IPCC report [1]. The REMO regional trend underlies the application of the RCM REMO driven by ECHAM5 [60], which is applied in several studies in this region [8,21,60–62]. Among other RCMs, REMO represents a moderate temperature and precipitation development [63,64]. Several RCMs underestimate for example the summer drying especially in the Danube region. REMO, however, shows a very modest bias [64]. To cover a range of uncertainties a second RCM trend, the MM5 regional trend, which is based on the application of the regional climate model MM5 and is also driven by ECHAM5 [65], was applied. The background of the Extrapolation trend is the analysis of temperature and precipitation trends of historic climate stations data of the German (DWD) and Austrian (ZAMG) weather services, analysed by Reiter et al. [66].

To show the characteristic development of the meteorological input data, Table 2 gives an overview of the temperature and precipitation changes between 1990 and 2100 regarding the four climate trends considering annual and semi-annual summer (May–October) and winter (November–April) periods. For the annual values all trends show a clear increase in temperature and a varying decline in precipitation until 2100. In general, the IPCC regional trend is the weakest, followed by the rather moderate trends MM5 regional and REMO regional. The Extrapolation trend is the most severe one. For all four trends temperature increase results in less annual precipitation with a clear decrease in summer and a slighter increase in winter. Thereby, comparing the middle trends, REMO regional shows a slightly higher decrease in summer and a smaller increase in winter than MM5 regional.

TABLE 2: Temperature increase (K) and precipitation changes (%) between 1990 and 2100 regarding the four climate trends IPCC regional, MM5 regional, REMO regional and Extrapolation.

Climate Trend	Temperature Increase (K)	Precipitation Changes (%)		
	Annual	Annual	Winter (November–April)	Summer (May–October)
IPCC regional	+3.3	−4.4	+2.1	−10.2
MM5 regional	+4.7	−3.5	+10.1	−14.6
REMO regional	+5.2	−12.6	+7.3	−16.6
Extrapolation	+5.2	−16.4	+9.0	−30.7

As shown in Section 4.1, the temperature and precipitation changes of each trend trigger in turn by a more or less extent, changes in runoff, the hydrological storage and thereafter also in hydroelectric power generation.

7.4 RESULTS AND DISCUSSION OF FUTURE SCENARIOS

The results, analysed in the following chapters, display a possible range of future development of the hydroelectric power generation under a changing meteorological and hydrological situation in the Upper Danube basin for the next 50 years. The applied climate change scenario conditions are based on the global IPCC-SRES-A1B emission scenario and consider uncertainties in the climate projections. To cover a near and a far future time period, the two decades 2021–2030 and 2051–2060 are regarded, which are compared with the reference decade 1991–2000. These three decades show future changes in equal time steps of 30 years. In general, the development of the hydroelectric power generation is highly correlated with the meteorological development of the specific climate trends shown in Table 2 and the triggered hydrological changes in runoff, evaporation, the snow and ice storage and the occurrence of the extreme events low-flows and floods addressed in Section 4.1. The mean annual and semi-annual summer and winter development of the hydroelectric power generation of all hydropower plants in the investigation area are shown in Section 4.2. To specify regional differences, the development of the mean monthly course of the hydroelectric power generation of three hydropower plants in hydrologically different parts of the investigation area are analysed in Section 4.3. Further, the future regional developments of the main runoff components rain, snow- and ice-melt and their impact on the hydroelectric power generation are discussed.

7.4.1 FUTURE METEOROLOGICAL AND HYDROLOGICAL DEVELOPMENT IN THE UPPER DANUBE BASIN

Besides a clear temperature increase, the water resources in the Upper Danube basin will be affected by climate change. Table 3 gives an overview on changes of important meteorological and hydrological variables for the four climate trends considering the two future decades 2021–2030 and 2051–2060 compared with the reference decade 1991–2000.

TABLE 3: Development of the meteorological and hydrological situation in the Upper Danube basin for the future decades 2021–2030 and 2051–2060 compared with the reference decade 1991–2000 considering the four climate trends IPCC regional, MM5 regional, REMO regional and Extrapolation.

Meteorological and hydrological variables	1991–2000	Climate trend	2021–2030		2051–2060	
			Av.	Δ	Av.	Δ
Mean annual air temperature (°C)	7.01	IPCC	8.04	+1.03	9.18	+2.17
		MM5	8.26	+1.25	9.70	+2.69
		REMO	8.43	+1.42	10.06	+3.05
		Extrapolation	8.08	+1.07	9.75	+2.74
Mean annual precipitation sum (mm)	1060	IPCC	1071	+1%	1053	−1%
		MM5	1052	−1%	1055	0%
		REMO	1020	−4%	1013	−4%
		Extrapolation	1061	0%	959	−10%
Mean annual snow precipitation fraction (%)	22	IPCC	21	−1	18	−5
		MM5	20	−2	17	−5
		REMO	21	−2	13	−10
		Extrapolation	22	0	16	−6
Mean annual evapotranspiration (mm)	416	IPCC	445	+7%	459	+10%
		MM5	415	0%	429	+3%
		REMO	419	+1%	442	+6%
		Extrapolation	426	+2%	433	+4%
Mean runoff at Achleiten outlet gauge (m^3/s)	1480	IPCC	1568	+6%	1408	−5%
		MM5	1482	0%	1444	−2%
		REMO	1397	−6%	1315	−11%
		Extrapolation	1546	+4%	1286	−13%
Water stored as glacier ice (10^6 m^3)	16,591	IPCC	3979	−76%	292	−98%
		MM5	4044	−76%	199	−99%
		REMO	3833	−77%	188	−99%
	Extrapolation	4157	−75%	214	−99%	
Average amount of snowmelt (mm)	332	IPCC	319	−4%	271	−18%
		MM5	306	−8%	259	−22%
		REMO	297	−11%	224	−33%
		Extrapolation	339	+2%	270	−19%

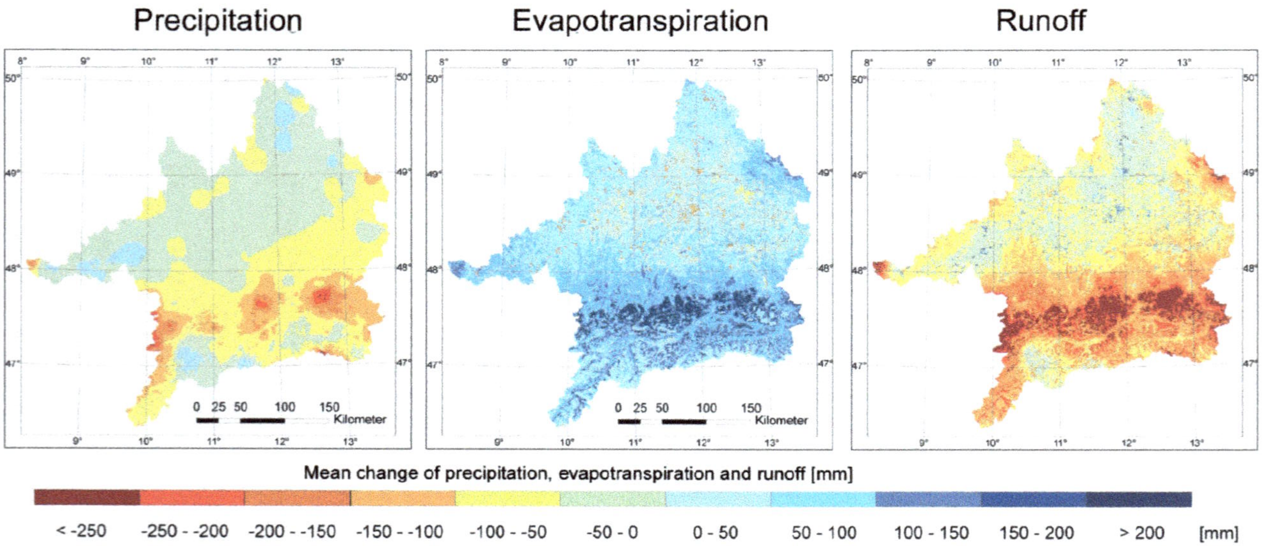

FIGURE 8: Difference of the water balance between the two periods 2036–2060 and 1971–2000 in the Upper Danube basin. This calculation is based on the trend REMO regional (based on Prasch and Mauser [37]).

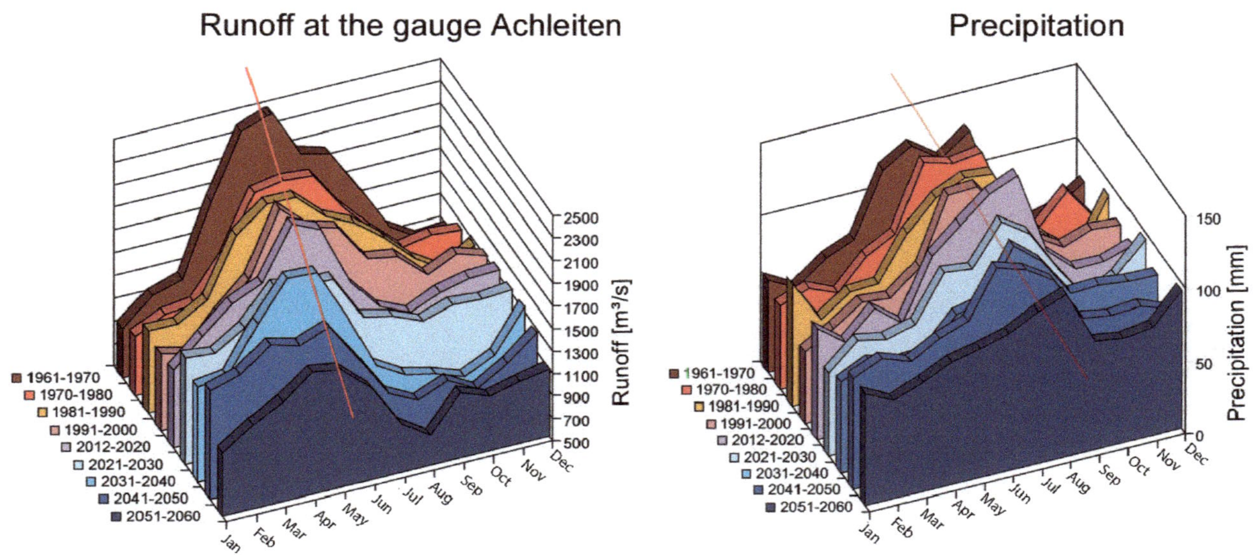

FIGURE 9: Decadal course of monthly precipitation in the Upper Danube basin and discharge from 1961 to 2060 at the Achleiten outlet gauge. This calculation is based on the trend REMO regional (based on Prasch and Mauser [37]).

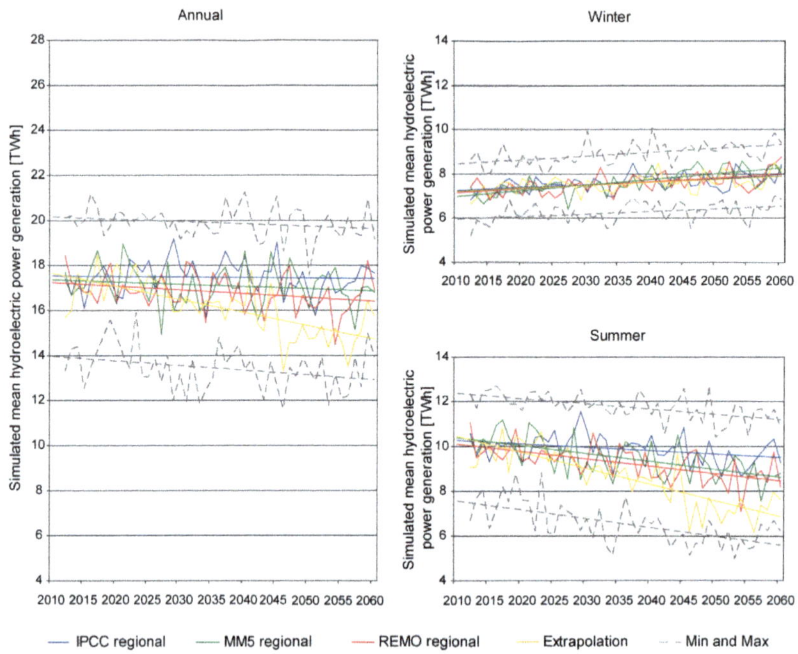

FIGURE 10: Development of hydroelectric power generation in the Upper Danube basin of annual and semi-annual summer and winter values for the time period 2011–2060 considering the means of the four climate trends IPCC regional, MM5 regional, REMO regional and Extrapolation and minima and maxima of all 16 climate scenarios.

For the mean of each decade Table 3 shows values for mean annual temperature and mean annual precipitation sums as well as values for the mean annual snow precipitation fraction, mean annual evapotranspiration, mean runoff at the Achleiten gauge, water stored as glaciers and the average amount of snowmelt. Depending on each climate trend, temperature increase causes a more or less severe rise in evapotranspiration, a negative trend in precipitation with a clear reduction of the snow precipitation fraction and a reduction of the mean annual runoff at the Achleiten gauge until 2060. According to the trend analysis based on Mann [67] and Kendall [68] the future trends of the annual runoff is significant for the average of REMO (significant on the 0.01 level of signification), MM5 (significant on the 0.001 level of signification) and Extrapolation (significant on the 0.1 level of signification), whereas for the IPCC scenario outputs no significant trend was detected. Together with the changes in precipitation, future water availability will be reduced particularly along the northern rim of the Alps, where precipitation will decrease and evapotranspiration increase because of a longer growing season as shown exemplarily for the rather medium climate trend REMO regional (see Figure 8). In contrast, in the northern part of the watershed a slight increase of water availability is partly modelled, which is due to small changes in the total amount of precipitation and a reduction of evapotranspiration.

Regarding the decades between 1961 and 2060, the mean seasonal course of runoff at the Achleiten outlet gauge indicates a shift of the mean monthly runoff peak from summer to spring, exemplarily shown for the climate trend REMO regional (see Figure 9). This shift is attributed to the reduction of the snow storage and an earlier snow-melt in the year as well as an increase in evapotranspiration and less precipitation during summer. However, the seasonal course of precipitation with its maximum during summer will flatten, but will not change remarkably [37].

The future low-flow conditions show a remarkable change in the low-flow regime of the Danube as well as severe changes in peak low-flow and its frequency of occurrence [32,69]. Low-flow will decrease distinctly under all trends at the Achleiten outlet gauge until 2060. However in contrast to a reduction along the Upper Danube River, an increase of low-flow is modelled in the alpine valleys. Reasons are a transformation of snowfall into rainfall and melting glaciers. Regarding the development of flood peaks, an increase is indicated in the alpine valleys and head-watersheds. In the other parts of the Upper Danube basin, the flood peaks stay almost stable. This again can be explained by changes of the mean annual snowfall fraction compared with total mean annual precipitation [37].

7.4.2 HYDROPOWER DEVELOPMENT IN THE UPPER DANUBE BASIN

The development of the hydroelectric power generation is visualized in Figure 10 for annual and semi-annual summer (May–October) and winter (November–April) values for the time period 2011–2060. To reduce uncertainties of single climate scenarios and to get a clearer picture about the development for each trend, the means of the model outputs of the four climate scenarios were taken. Further, to cover a plausible range of all 16 climate scenarios, the minima and maxima values of the climate scenarios were considered. Table 4 complements the analysis by showing explicit values of the development in percent for the future decades 2021–2030 and 2051–2060 based on the reference period 1991–2000. For the past decade, the total mean annual hydroelectric power generation of all considered hydropower plants in the Upper Danube basin reached 17.6 TWh, whereof about 70% were produced through runoff-river power plants and about 30% through reservoir hydropower plants. The energy production for the mean summer values accounted to 11.1 TWh and for the mean winter values 6.5 TWh.

7.4.2.1 DEVELOPMENT OF ANNUAL PRODUCTION

Regarding the development of the mean annual hydroelectric power generation in the Upper Danube basin in Figure 10, a general decline was determined for all climate trends and the minima and maxima values until 2060. Thereby, Table 4 shows that the IPCC regional trend is rather moderate, with a decline of 2% until the future decade 2051–2060. However, for the nearer future decade 2021–2030 a slight increase of 2% was modelled. MM5 regional shows a slight decrease of 3% for the first future decade and of 4% for the second future decade. Hydroelectric power generation simulated with REMO regional can be classified in the middle range with a decline by 5% for the first future decade and 7% for the second decade. Extrapolation displays the most severe values with a decrease of 15% until 2060, but indicates only a slight decrease of 3% until 2030. In general, under these climate trends the mean annual hydroelectric power generation will decline to a range of 17 to 18 TWh in 2021–2030, respectively of 15 to 17 TWh in 2051–2060. These results show a close correlation with the development of the precipitation situation shown in Section 3.4 for each trend, whereby IPCC regional shows the lowest annual decrease, followed by MM5 regional and REMO regional and the severest annual decline with Extrapolation. Besides the decrease in precipitation, the decline of the mean annual hydroelectric power generation is caused mostly by an increase in mean annual evapotranspiration and a decrease in mean annual runoff as shown in Section 4.1.

TABLE 4. Development of hydroelectric power generation in the Upper Danube basin by annual and semi-annual summer and winter values for the future decades 2021–2030 and 2051–2060 compared with the reference decade 1991–2000 considering the four climate trends IPCC regional, MM5 regional, REMO regional and Extrapolation.

		1991–2000	2021–2030		2051–2060	
		(TWh)	(TWh)	(%)	(TWh)	(%)
Annual	Reference	17.6	–	–	–	–
	IPCC regional	–	17.9	+1.5	17.2	−2.2
	MM5 regional	–	17.2	−2.5	16.9	−3.8
	REMO regional	–	16.8	−4.7	16.4	−6.7
	Extrapolation	–	17.1	−3.0	15.0	−15.0
Summer	Reference	11.1	–	–	–	–
	IPCC regional		10.4	−6.4	9.4	−15.3
	MM5 regional		9.8	−11.7	8.8	−20.5
	REMO regional		9.4	−15.1	8.5	−23.0
	Extrapolation	9.6	−13.5	7.1	−35.6	
Winter	Reference	6.5	–	–	–	–
	IPCC regional	–	7.5	+15.0	7.8	+19.9
	MM5 regional	–	7.4	+12.3	8.1	+24.5
	REMO regional	–	7.4	+12.7	7.9	+21.0
	Extrapolation	–	7.5	+14.7	7.8	+19.8

7.4.2.2 DEVELOPMENT IN WINTER

Currently the winter energy production in the Upper Danube basin is lower than in summer. However, as illustrated in Figure 10, all four climate trends show an increase of the hydroelectric power generation in the Upper Danube basin for the winter periods until 2060. The main reason for this development is the general increase of precipitation in winter (Section 3.4) and a decline in solid precipitation and thereby snow storage, leading to more runoff. As a consequence, the future alpine low-flow situation in winter will be less severe leading in turn to a further increase of hydroelectric power generation. Furthermore, the range of the minima und maxima energy production in winter is lower indicating an inferior natural variability than in summer. For both future decades, all four trends show, as listed in Table 4, a similar increase by 20% (IPCC regional and Extrapolation), 21% (REMO regional) and 25% (MM5 regional) until 2060; whereby the latter originates in the highest increase in winter precipitation. In general, compared with the 6.5 TWh of the reference decade, the hydroelectric power generation for all hydropower plants is said to increase in winter between 7.4 to 7.5 TWh in the decade 2021–2030, and 7.8 to 8.1 TWh in the decade 2051–2060. This development leads to a significant increase of hydroelectric power generation during winter.

7.4.2.3 DEVELOPMENT IN SUMMER

Contrary to the winter development, for the summer periods, a decline of hydroelectric power generation in the Upper Danube basin is displayed in Figure 10 for all trends until 2060. This is mainly triggered by a decrease in summer precipitation as shown in Section 3.4, a decline in snow- and ice-melting rates and an increase in evapotranspiration. Moreover, low-flow will decrease remarkably under all trends at the Achleiten outlet gauge until 2060, leading to more losses in hydroelectric power generation. Besides, the minima and maxima values delineate a big range of possible summer values of the single climate scenarios, indicating a high natural variability. The IPCC regional trend is again rather moderate, the trends

MM5 regional and REMO regional lie in the middle and the trend Extrapolation at the upper boundary. Until 2060 the trends decline with a range between 15% (IPCC regional) and 36% (Extrapolation) (see Table 4). The more moderate trends MM5 regional and REMO regional indicate a decrease by 21 and 23%, respectively. As an exception, during 2021–2030, the results of the trend REMO regional is slightly more severe than of the trend Extrapolation. The reason for this is that the summer decline in precipitation does not increase linearly. For the trend Extrapolation, e.g., the development is more pronounced in the second scenario period. Whereas the summer energy generation was 11.1 TWh in the reference decade, the hydroelectric power generation for all hydropower plants in the investigation area declines from 9.4 to 10.4 TWh in the decade 2021–2030, and from 7.1 to 9.4 TWh in the decade 2051–2060. This indicates obvious future energy production losses during summer.

7.4.2.4 DEVELOPMENT OF THE SUMMER AND WINTER FRACTION

Table 5 lists the development of the summer and winter fractions of hydroelectric power generation in the Upper Danube basin for the reference decade 1991–2000 and the two future decades 2021–2030 and 2051–2060. In 1991–2000 the ratio of summer to winter production was 63 to 37%. Whilst the past energy production was considerable higher in summer than in winter, the summer and winter fractions become more inter-annually equalized until 2060. In 2021–2030 the ratio shows already a range of 56 to 58% for the summer production and 42 to 44% for the winter production depending on the climate trends. At the end of the simulation period the values of the summer and winter periods will be quite similar with summer production percentages of 47 to 55% and winter percentages of 45 to 53% depending on the climate trends. The trend Extrapolation even indicates a higher winter than summer production for the decade 2051–2060. This means that the semi-annual hydroelectric power generation will be more balanced in the future.

In general, the future development of the mean annual hydroelectric power generation of all hydropower plants in the Upper Danube basin indicates a decline with its severity depending on the respective climate trend. Whilst the summer values experience a decrease with a large range of minima and maxima values of the single scenarios, the winter values increase with a low variability until 2060.

TABLE 5: Development of the mean semi-annual summer and winter energy production fraction (%) based on the mean annual hydroelectric power generation in the Upper Danube basin for the decades 1991–2000, 2021–2030 and 2051–2060 considering the four climate trends IPCC regional, MM5 regional, REMO regional and Extrapolation.

	1991–2000		2021–2030		2051–2060	
	Summer	Winter	Summer	Winter	Summer	Winter
Reference	63.1	36.9				
IPCC regional			58.1	41.9	54.7	45.3
MM5 regional			57.0	43.0	52.1	47.9
REMO regional			56.0	44.0	51.8	48.2
Extrapolation			56.1	43.9	47.3	52.7

7.4.3 REGIONAL DIFFERENCES WITHIN THE UPPER DANUBE BASIN

In this Section different regional developments within the Upper Danube basin are analysed. To work out such regional differences, three hydropower plants were chosen, situated in hydrologically different parts, which diverge in their degree of alpine character: Donauwoerth (Danube), Wasserburg (Inn) and Kaunertal (Gepatsch reservoir) (see pink circles in Figure 2). After a short description of their main hydrological and hydroelectric characteristics, the past and future contributions of the runoff components rain, snow- and ice-melt are shown. This leads to a discussion on their regional differences and their inter-annual development. Hence, the development of the three demonstrated hydropower plants should also represent the development of others with similar regional characteristics.

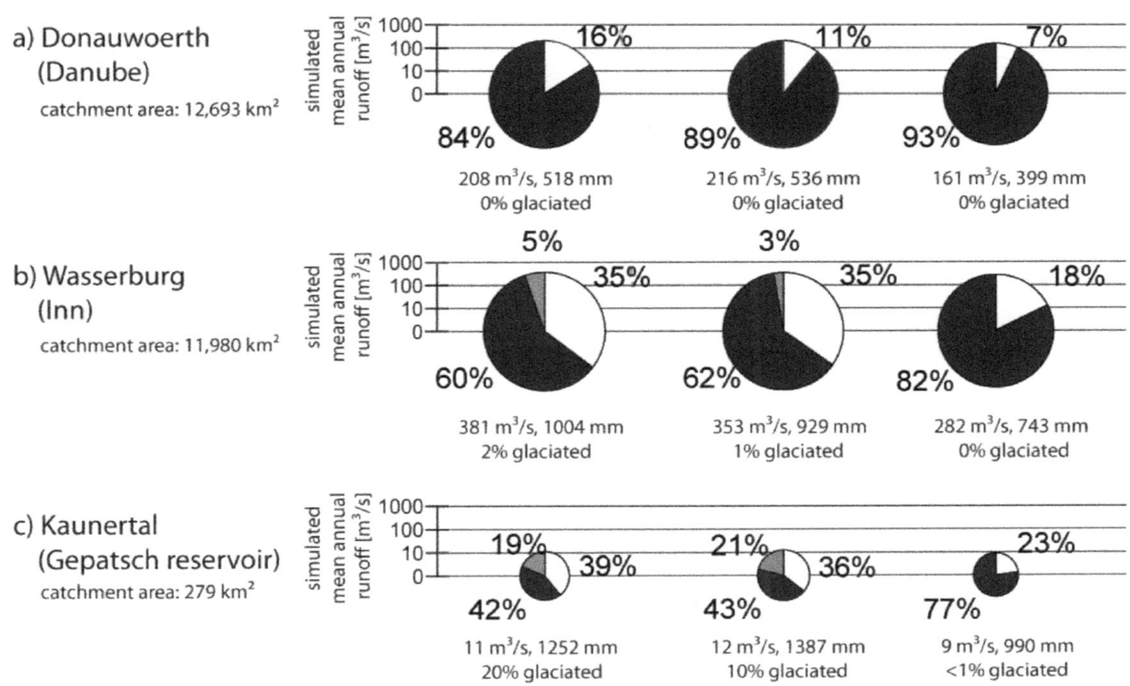

FIGURE 11: Contribution of the runoff components rain (dark gray), snow- (light gray) and ice-melt (white) at the hydropower plants (a) Donauwoerth, (b) Wasserburg and (c) Kaunertal for the decades 1991–2000, 2021–2030 and 2051–2060. The future decades are simulated under the terms of a medium climate scenario of the climate trend REMO regional. Total mean annual runoff values (mm and m^3/s) and the glaciated area (%) are listed below the pie charts.

Table 6. Main hydroelectric characteristics of the hydropower plants (a) Donauwoerth (Danube), (b) Wasserburg (Inn) and (c) Kaunertal (Gepatsch reservoir).

	(a) Donauwoerth	(b) Wasserburg	(c) Kaunertal
Location at river/reservoir	river Danube	river Inn	Gepatsch reservoir, filled by the stream Faggenbach and several water transfers
Geogr. coordinates	E 10.77°, N 48.70°	E 12.22°, N 48.06°	E 10.74°, N 46.96°
Type of hydropower plant	runoff-river power plant	runoff-river power plant	reservoir power plant
Start year of operation	1984	1938	1964
Mean hydraulic head	5.5 m	7.0 m	844.0 m
Maximum turbine discharge	200 m^3/s	465 m^3/s *	54 m^3/s
Maximum capacity	8.5 MW	24 MW *	370 MW
Mean annual hydroelectric power generation	54.8 GWh	148.3 GWh *	661.0 GWh
Data base	[70,71]	[70,72]	[73]

* *In 2009 the maximum turbine discharge and the maximum capacity of the hydropower plant Wasserburg (Inn) was increased by installing a new turbine in one section. This achieved an additional mean annual hydroelectric power generation of 12.6 GWh [72]. However, this change was not considered for the future simulations. So the same constellation than in the reference period 1991–2000 was assumed for a better comparison.*

7.4.3.1 HYDROLOGICAL AND HYDROELECTRIC CHARACTERISTICS

The three hydropower plants cover typical hydrological differences with more or less alpine influence. Table 6 gives an overview about the main hydroelectric characteristics.

The three facilities are briefly described hereafter:

1. The Donauwoerth runoff-river power plant is situated at the upper part of the river Danube with a catchment area of 12,693 km^2. It is dominated by the mid-altitude mountains in the Northwest and has

little alpine influence from the tributary Iller in the South. Therefore the present runoff regime is predominantly characterised by rain (pluvial). Donauwoerth is a medium sized runoff-river power plant with a mean annual hydroelectric power generation of approx. 60 GWh and a maximum turbine discharge of 200 m^3/s.

2. The Wasserburg runoff-river power plant at the lower part of the river Inn with a catchment area of 11,980 km^2 is mostly influenced by the snow storage and also partly by the ice storage of the Central Alps. The present runoff regime is predominantly influenced by snow-melt (nival). Wasserburg is a large runoff-river power plant with a mean annual hydroelectric power generation of approx. 150 GWh and a maximum turbine discharge of approx. 500 m^3/s.
3. The Kaunertal reservoir hydropower plant has a small catchment area of 279 km^2 and is situated in a highly glaciated alpine head-watershed in the Austrian Central Alps. Therefore the present runoff regime is predominantly composed by the snow and ice storage (glacio-nival). With a mean annual hydroelectric power generation of 661 GWh, the power plant has a maximum turbine discharge of 54 m^3/s and a mean hydraulic head of 844 m, depending on the filling line of the Gepatsch reservoir. The reservoir outflow and storage volume is determined using a standard monthly operation plan (see Figure 3).

7.4.3.2 CONTRIBUTION OF THE RUNOFF COMPONENTS RAIN, SNOW- AND ICE-MELT

The regional hydrological differences of the three hydropower plants can be described, besides different regional precipitation patterns, by the contribution of the runoff components rain, snow- and ice-melt. Prasch [39] and Weber et al. [10] presented a new technique to model the runoff components rain, snow- and ice-melt for bigger watersheds with the hydrological model PROMET. By different model run settings, the snow- and ice-melt as well as the liquid fraction of precipitation can be calculated quantitatively for each grid cell. Ice-melt denotes the water released from the ice body of the snow-free glacier surface, which is modelled in detail by the energy-mass-balance model SURGES [11,39], snow-melt denotes the water that contributes to runoff from snow-melt in the watershed [9]. For the future simulations one of the four climate scenarios of the climate trend REMO regional was chosen exemplarily as a medium climate scenario.

Regarding the reference decade, the Wasserburg hydropower plant receives with 381 m^3/s the highest mean annual runoff (Figure 11). Although the catchment area of Donauwoerth is larger than that of Wasserburg, this hydropower plant only receives 208 m^3/s. The mean annual fill rate of the Gepatsch reservoir is 11 m^3/s. At Donauwoerth, the mean annual runoff is supplied mostly by rain (84%) with a smaller proportion of snow-melt (16%). Compared with Donauwoerth, the snow-melt runoff contribution (35%) of the more alpine influenced Wasserburg site is considerably higher. There, the runoff is triggered in addition by a small proportion of ice-melt (5%) due to its glaciated area of 2%. The Kaunertal site, with its 20% glaciated head-watershed, has a large fraction of snow- (39%) and ice-melt (19%). However, the rain contribution is highest at all sites.

In the future, all three hydropower plants experience a decline of the mean annual runoff, in particular in the second future decade, because of a considerable decrease in water availability due to a decrease in the mean annual precipitation and an increase in evapotranspiration. Until 2060, the mean annual runoff accounts to 161 m^3/s at Donauwoerth, to 282 m^3/s at Wasserburg and to 9 m^3/s at Kaunertal for the chosen climate scenario. Figure 11 shows the mean annual runoff components runoff, snow- and ice-melt for the reference and the two future decades. Whilst the contribution of rain increases in future, the contribution of snow-melt declines for all locations. Especially in the second scenario decade, the snow-melt decreases considerably to 7% at Donauwoerth, to 18% at Wasserburg and to 23% at Kaunertal. Even at the highly alpine location Kaunertal the decrease in snow-melt is strong, which results from the decrease of the snow storage even in high altitudes. At Wasserburg, the small proportion of mean annual ice-melt in the past becomes even smaller in 2021–2030 (3%) because of an obviously higher contribution of the rain component. However, at Kaunertal the runoff component ice-melt increases slightly in 2021–2030 (21%) due to an increased glacier melt. Because of a progressive glacier retreat, most of the glaciers will be extinct in the Eastern Alps until 2060 [10] with a total glacier coverage of less than 1% as shown in Figure 11. Therefore no ice-melt was simulated in 2051–2060.

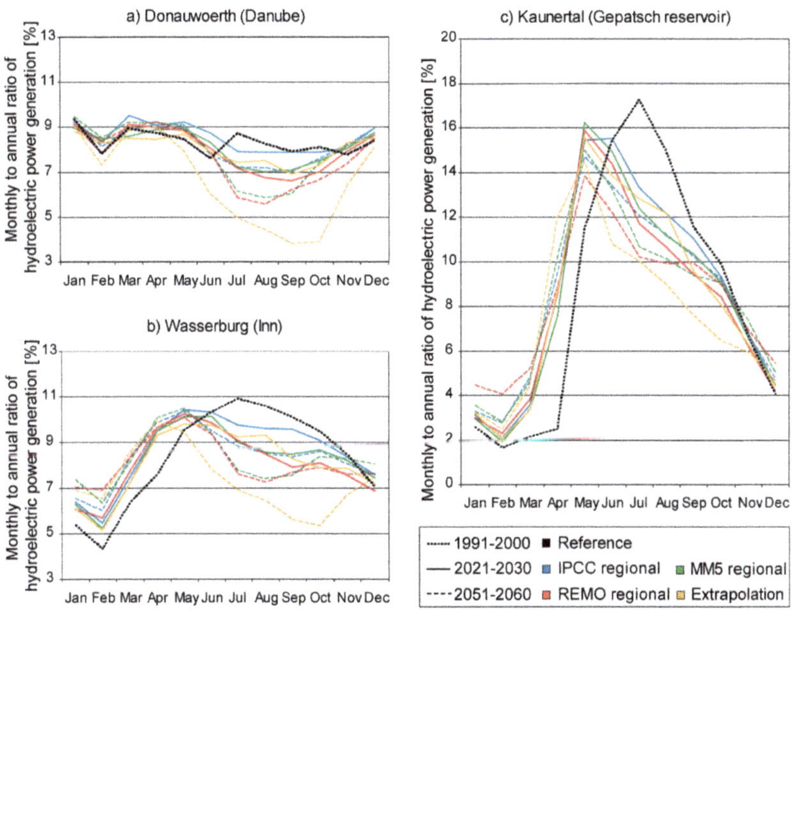

FIGURE 12: Development of mean monthly hydroelectric power generation at the hydropower plants (a) Donauwoerth, (b) Wasserburg and (c) Kaunertal for the reference decade 1991–2000 and the future decades 2021–2030 and 2051–2060 considering the climate trends IPCC regional, MM5 regional, REMO regional and Extrapolation.

Overall, the results of the medium range climate scenario indicate for all three locations a clear quantitative future change of the mean annual runoff components, with an increase of the rain-component and a decrease of the components snow- and ice-melt. The other climate scenarios would show a similar development with more or less severe characteristics. Besides these quantitative changes of the mean annual runoff and its components, seasonal changes will occur [11]. The inter-annual courses of the compartment rain will shift due to changes in precipitation seasonality. The inter-annual courses of snow- and ice-melt, however, will shift towards an earlier outset of melting within the year. The reasons for the shift in snow-melt are a general reduction of snow fall compared to rain fall in winter and due to higher temperatures triggering earlier melting conditions in the year. The shift in ice melt is triggered mostly because glaciers are snow-free much earlier [11]. These effects cause different inter-annual runoff and consequently hydroelectric power generation patterns in the future, which will be shown for the three hydropower plants in the following.

7.4.3.3 DEVELOPMENT OF HYDROELECTRIC POWER GENERATION

Figure 12 illustrates the mean annual course of hydroelectric power generation of the Donauwoerth, Wasserburg and Kaunertal power plants for the decades 1991–2000, 2021–2030 and 2051–2060.

The curves are based on the ratio of monthly versus annual hydroelectric power generation for each decade for the four climate trends IPCC regional, MM5 regional, REMO regional and Extrapolation. The regional developments at the hydropower plants (a) Donauwoerth Danube), (b) Wasserburg (Inn) and Kaunertal (Gepatsch reservoir) are discussed below:

1. During the reference period the mean annual course of hydroelectric power generation at Donauwoerth is generally balanced with small peaks in January, March and July. In the past, the variability was very low; each month had a mean fraction of 8 to 9% of the mean total annual production. In the future, depending on the intensity of the respective climate trend, the production declines in

the summer and autumn months and increases slightly in the winter months, leading to an increase of the monthly variability with a range from 4 to 9.5% of the mean total annual production considering all climate trends. Whilst the development for all trends is less pronounced until 2030 and for the trend IPCC regional also until 2060, due to a large decline in summer precipitation Extrapolation shows an extraordinarily severe decline in the second scenario period with distinct minima in September and October, whereas the trends MM5 regional and REMO regional indicate the future minimum in August. Because the runoff component rain has the highest influence at this location, the changes are predominately triggered by changes in precipitation with a distinct decrease in summer and a marked increase in winter. The future decrease of the runoff component snow-melt plays only a minor role because its contribution is generally very low. The development of hydroelectric power generation at this site is mainly influenced by changes in runoff variability. However, low-flow events become more severe in summer at less alpine influenced sites, leading to possible production restrictions, especially during the second simulation period.

2. Regarding Wasserburg, the graph of hydroelectric power generation in the past shows a distinct annual cycle with a minimum in February due to the highest snow storage during this month, and its maximum in July depending on snow- and ice-melt and a the summer precipitation peak. The inter-annual variability is high with mean monthly fractions of 4.5 to 11% of the mean total annual production. The trends show a shift of the former July-maximum towards April and May, whilst the summer months with their former high production values tend to reach a clear minimum especially under the climate trend Extrapolation in the second scenario decade. This is triggered by a general reduction of the snow-melt and due to warmer temperatures to an earlier outset of the snow-melting season in spring. However, the former minimum in February still remains, because during this month the snow storage is still highest. But the reduction of the February-minimum indicates a higher winter rain contribution and a general decrease of the snow storage. Moreover, less low-flow occurs in winter, which

leads to fewer restrictions in hydroelectric power generation. The main reason for this is the transformation of snowfall into rainfall in winter leading to more runoff in this time period. In contrast to Donauwoerth, the inter-annual variability experiences a slight decline with a future range of mean monthly fractions of 5.5 to 10.5% until 2060, because the future mean minima and maxima values are less pronounced than during the reference period. Summarized, at this location the inter-annual changes are mostly triggered by changes of the snow storage with an increasing influence of the runoff component rain. Thus, the development of the hydroelectric power generation will be more and more triggered by the runoff component rain.

3. As described in Section 3.1, a standard monthly-based operation plan for the Gepatsch reservoir of the Kaunertal power plant is included, which shifts the natural annual runoff course due to assumed electricity demands. The management plan refers to present conditions and is not changed for the future simulations because of an uncertain development in energy strategies and demand. However, only monthly values were assumed, not taking into account detailed management structures. Due to no further detailed information about a higher temporal resolution, this assumption should be seen as a reasonable approach. The past annual course of the hydroelectric power generation has, because of its highly alpine character, a greater variability than Wasserburg or Donauwoerth with mean monthly fractions of 2 to 17% of the mean total annual production. Although the implemented management plan shifts the reservoir inflows temporarily, in August a clear maximum occurred due to large snow- and ice-melting rates, whereas in February a clear minimum due to high winter snow storage, was observed. In the future, the development of the four climate trends is similar to Wasserburg, however, with a drastic decline in ice-melt influence until 2060. The monthly production variability decreases slightly during the decades 2021–2030 and 2051–2060 because of less pronounced mean minima and maxima values. But there still remains a high variability of 2.5 to 15% until 2060. Like at the Wasserburg location, the reduction of the February-minimum indicates

a higher winter rain contribution for the future, especially for the trend MM5 regional. Because of earlier snow-melt, glacier ice will become snow-free sooner in the year, which increases glacier melting. In 2051–2060 the highest energy production occurs in May for all four climate trends. Hydroelectric power generation will still be high during summer, but will experience a strong decline due to a decrease of the snow storage and glacier retreat resulting in less snow- and ice-melt, especially in the second future decade. This summer development is especially pronounced for the trend Extrapolation.

To summarize, the three hydropower plants show different annual courses of hydroelectric power generation for the reference and the two future decades. Besides slightly different precipitation patterns, the differences originate in the dissimilar contributions of the runoff components rain, snow- and ice-melt. Changes at less alpine sites (e.g., Donauwoerth) are predominantly influenced by rain trends and lead to an increase in annual variability with a considerable minimum in summer to autumn. The development at alpine sites (Wasserburg and Kaunertal), reflect furthermore changes in the snow storage and for glaciated headwaters also in the ice storage leading to a slight decrease in annual variability and a shift of the maximum production from summer to spring.

7.5 CONCLUSIONS

In order to face and overcome future development tendencies, different adaptation measures could be considered for hydropower planning strategies. Thereby quantitative, seasonal and regional tendencies, as analysed in this study, should be taken into account, such as the altered seasonal rain patterns and the influence of the timing of the snow- and ice-melt on the hydroelectric power generation. To cover the growing energy demand, especially in the green energy sector, many regions already started to release feasibility studies to build new facilities and to enhance, reactivate and modernize old power plants [74,75]. However, all planning strategies should include possible trends considering climate change conditions.

This is also the case when considering, e.g., future energy market strategies and possibilities for the future composition of the energy mix.

For the Upper Danube basin hydroelectric power generation will be considerably influenced by regional climate change. Future precipitation changes and temperature increase will trigger changes in the hydrological cycle. These effects lead in turn to changes of the energy production as shown in this study for 16 climate scenarios underlying four climate trends based on the global IPCC-SRES-A1B scenario. The general trend of hydroelectric power generation for all hydropower plants in the Upper Danube watershed until 2060, as analysed in Section 4.2, indicates a decline with its severity depending on the selected climate trends. Whilst the summer values experience a decrease with a large annual variability within each scenario, the winter values increase with a low variability until 2060. Moreover, changes in the runoff components rain, snow- and ice-melt trigger seasonal and regional changes in hydroelectric power generation as shown in Section 4.3. Less alpine sites, which are predominantly influenced by the runoff component rain, show an increase in future annual variability with a considerable minimum in summer to autumn due to less future precipitation during this season. At alpine sites, the presently quite large proportion of the runoff component snow-melt and in glaciated headwaters also the ice-melt decreases leading to a slight decrease in annual variability and a shift of the peak production from summer to spring. Due to a larger contribution of the runoff component rain and less influence of the component snow- and ice-melt for the whole Upper Danube, the seasonal patterns become more similar in all parts.

To give an outlook and to make some critical reflections, the following points should be considered. This modelling approach shows a plausible future development under climate change conditions with the advantage of taking into account seasonality, specific runoff components and the development of single hydropower plants and thus the possibility of considering regional differences. However, climate projections should be seen as scenarios and the model technique as a plausible representation of physical conditions. Firstly, climate change can be simulated by different scenarios indicating varying developments, which should be seen as a range of possible future developments. But, when considering the simulated results, uncertainties of the global emission scenarios and the regional climate assumptions should be taken into account. For this study, the medium ranged, quite probable global IPCC-SRES-A1B emission scenario [1] was chosen for the 16 climate scenarios based on four regional climate trends. The resulting simulations show a plausible range of future development of energy production in the Upper Danube basin. However, further assumptions on emission scenarios or regional climate trends as well as the consideration of different time periods can lead to a larger or smaller extent of future changes. Moreover, there are still more uncertainties in the application of hydrological modelling approaches in climate change scenario simulations as for instance, besides the uncertainty in the future greenhouse gas emissions, there are also uncertainties in the resulting changes in climate, changes in water management practices like irrigation and water supply, land use changes, impacts of economic change, population development and many more. Accordingly, these uncertainties in the results of scenario studies should always be considered. Secondly, with the fully distributed hydrological model PROMET and its coupled hydropower module used in this study, each runoff-river and reservoir power plant was considered individually taking into account the advantage to calculate the capacity of each hydropower plant with its individual parameterization, instead of transferring future runoff projections to the future hydropower situation, as shown in other studies. Regarding the simulated mean annual hydroelectric power generation they highly match the actual validated values. In addition the short term validation of daily generation at runoff-river power plants proves the capabilities of PROMET to reproduce the dynamic of runoff and power generation. However, for a daily or hourly time step more information of each individual power plant would be necessary. As hydroelectric power generation is influenced by extreme low-flow and flood events, e.g., due to restrictions, further research on this topic is certainly of interest. Although PROMET performs very well in generating reasonable runoff and hydroelectric power generation values, there are still some modelling uncertainties, e.g., displaying extreme flood events [32]. Moreover, due to a lack of available information, the implemented monthly-based reservoir operation plans, referred to present conditions, are only based on best guess assumptions. Further knowledge, e.g., provided by hydropower plant operators about the operation rules, the dependence of reservoir operation on short term electricity prices and a finer temporal resolution should be considered in the simulations and would consequently improve the results. In the future, the implemented reservoir management scenarios will certainly change due to changing energy markets, energy demand and reservoir functions. These changes however, are not considered in this study, but should be addressed in further research. Additionally, it would certainly be interesting to develop adaptation strategies as shown, e.g., in Payne et al. [26] to mitigate reservoir system performance losses due to climate change by examining several alternative reservoir operating policies.

REFERENCES

1. IPCC. Climate Change 2007: The Physical Science Basis; Contribution of Working Group I to the Forth Assessment Report of the Intergovernmental Panel on Climate Change; IPCC: Cambridge, UK, 2007.
2. REN21 (Renewable Energy Policy Network for the 21st Century). Renewables 2010. Global Status Report; REN21 Secretary: Paris, France, 2010.
3. Energiestatus Österreich. Bundesministerium für Wirtschaft, Familie und Jugend: Vienna, Austria. 2011. Available online: http://www.bmwfj.gv.at/EnergieUndBergbau/Energieversorgung/ Documents/A4_Kern_doppelseitig%20gross%20%28korr%29.pdf (accessed on 18 August 2011).
4. Wasserkraftnutzung. Fakten und Zahlen. Bundesamt für Energie: Bern, Switzerland, 2010. Available online: http://www.bfe.admin.ch/themen/00490/00491/index.html?lang=de&dossier_id=00803 (accessed on 18 August 2011).
5. Schaefli, B.; Hingray, B.; Musy, A. Climate change and hydropower production in the Swiss Alps: Quantification of potential impacts and related modelling. Hydrol. Earth Syst. Sci. 2007, 11, 1191–1205.
6. Beniston, M.; Stoffel, M.; Hill, M. Impacts of climate change on water and natural hazards in the Alps: Can current water governance cope with future challenges? Examples from the European "ACQWA" project. Environ. Sci. Policy 2011, in press.
7. Blöschl, G.; Schöner, W.; Kroiß, H.; Blaschke, A.P.; Böhm, R.; Haslinger, K.; Kreuzinger, N.; Merz, R.; Parajka, J.; Salinas, J.L.; et al. Anpassungsstrategien an den Klimawandel für Österreichs Wasserwirtschaf—Ziele und Schlussfolgerungen der Studie für Bund und Länder. Oesterr. Wasser-Abfallwirtsch. 2011, 1–2, 1–20.
8. Klein, B.; Krahe, P., Lingemann, I.; Nilson, E.; Kling, H.; Fuchs, M. Assessing Climate Change Impacts on Water Balance in the Upper Danube Basin based on a 23 Member RCM Ensemble. In Proceedings of the XXVth Conference of the Danubian Countries, Budapest, Hungary, 16–17 June 2011.
9. Dobler, C.; Stötter, J.; Schöberl, F. Assessment of climate change impacts on the hydrology of the Lech Valley in northern Alps. J. Water Clim. Change 2010, 1, 207–218.

10. Weber, M.; Braun, L.; Mauser, W.; Prasch, M. Contribution of rain, snow- and icemelt in the Upper Danube discharge today and in the future. Geogr. Fis. Dinam. Quat. 2010, 33, 221–230.
11. Weber, W.; Braun, L.; Mauser, W.; Prasch, M. Die Bedeutung der Gletscherschmelze für den Abfluss der Donau gegenwärtig und in der Zukunft. Mitt. Hydrogr. Dienstes Österreich 2009, 86, 1–29.
12. Weingartner, R., Viviroli, D.; Schädler, B. Water resources in mountain regions: A methodological approach to assess the water balance in a highlandlowland-system. Hydrol. Process. 2007, 21, 578–585.
13. Horton, P.; Schaefli, B.; Mezghani, A.; Hingray, B.; Musy, A. Assessment of climate-change impacts on alpine discharge regimes with climate model uncertainty. Hydrol. Process. 2006, 20, 2091–2109.
14. Beniston, M. Mountain climates and climatic change: An overview of processes focusing on the European Alps. Pure Appl. Geophys. 2005, 162, 1587–1606.
15. Zierl, B.; Bugmann, H. Global change impacts on hydrological processes in Alpine catchments. Water Resour. Res. 2005, 41, 1–13.
16. Beniston, M. Climatic change in mountain regions: A review of possible impacts. Clim. Change 2003, 59, 5–31.
17. Viviroli, D.; Weingartner, R.; Messerli, B. Assessing the hydrological significance of the world's mountains. Mt. Res. Dev. 2003, 23, 32–40.
18. Arnell, N.W. The effect of climate change on hydrological regimes in Europe: A continental perspective. Glob. Environ. Change 1999, 9, 5–23.
19. Huss, M.; Jouvet, G.; Farinotti, D.; Bauer, A. Future high-mountain hydrology: A new parameterization of glacier retreat. Hydrol. Earth Syst. Sci. 2010, 14, 815–829.
20. Braun, L.; Weber, M.; Schulz, M. Consequences of climate change for runoff from Alpine regions. Ann. Glaciol. 2000, 31, 19–25.
21. Stanzel, P.; Nachtnebel, H.P. Mögliche Auswirkungen des Klimawandels auf den Wasserhaushalt und die Wasserkraftnutzung in Österreich. Oesterr. Wasser-Abfallwirtsch. 2010, 62/9–10, 180–187.
22. Kuhn, M.; Olefs, M. Auswirkungen von Klimaänderungen auf das Abflussverhalten von vergletscherten Einzugsgebieten im Hinblick auf die Speicherkraftwerke. In StartClim2007E; Institute of Meteorology and Geophysics, University of Innsbruck: Vienna, Austria, 2007.
23. Lehner, B.; Czisch, G.; Vassolo, S. The impact of global change on hydropower potential of Europe: A model-based analysis. Energy Policy 2005, 33, 839–855.
24. Piot, M. Auswirkungen der klimaerwärmung auf die wasserkraftproduktion der schweiz. Wasser Energie Luft 2005, 97, 365–367.
25. Madani, K.; Lund, J.R. Estimated impacts of climate warming on California's high-elevation hydropower. Clim. Change 2010, 102, 521–538.
26. Payne, J.T.; Wood, A.W.; Hamlet, A.F.; Palmer, R.N.; Lettenmaier, D.P. Mitigating the effects of climate change on water resources of the Columbia river basin. Clim. Change 2004, 62, 233–256.
27. Vicuna, S.; Dracup, J.A. The evolution of climate change impact studies on hydrology and water resources in California. Clim. Change 2007, 82, 327–350.

28. Christensen, N.; Wood, A.W.; Voisan, N.; Lettenmaier, D.P.; Palmer, N.R. The effect of climate change on the hydrology and water resources of the Colorado River basin. Clim. Change 2004, 62, 337–363.
29. GLOWA-Danube Projekt. Global Change Atlas Einzugsgebiet Obere Donau, 6th ed.; LMU Munich, Department of Geography: Munich, Germany, 2010. Available online: http://www.glowa-danube.de (accessed on 27 September 2011).
30. Mauser, W.; Ludwig, R. GLOWA-Danube—A Research Concept to Develop Integrative Techniques, Scenarios and Strategies Regarding Global Changes of the Water Cycle. In Climate Change: Implications for the Hydrological Cycle and for Water Management, Advances in Global Research; Beniston, M., Ed.; Kluwer Academic Publishers: Dordrecht, The Netherlands and Boston, MA, USA, 2002; Volume 10, pp. 171–188.
31. Ludwig, R.; Mauser, W.; Niemeyer, S.; Colgan, A.; Stolz, R.; Escher-Vetter, H.; Kuhn, M.; Reichstein, M.; Tenhunen, J.; Kraus, A.; et al. Web-based modelling of energy, water and matter fluxes to support decision making in mesoscale catchments—The integrative perspective of GLOWA Danube. Phys. Chem. Earth 2003, 28, 621–634.
32. Mauser, W.; Bach, H. PROMET—Large scale distributed hydrological modelling to study the impact of climate change of the water flows of mountain watersheds. J. Hydrol. 2009, 376, 362–377.
33. Prasch, M.; Bernhardt, M.; Weber, M.; Strasser, U.; Mauser, W. Physically based Modelling of Snow Cover Dynamics in Alpine Regions. In Proceedings of the Innsbruck Conference, Innsbruck, Austria, 15–17 October 2007; Volume 2, pp. 323–330.
34. Amt der Tiroler Landesregierung, Abteilung Wasser-, Forst- und Energierecht. Tiroler Energiestrategie 2020. Grundlage für die Tiroler Energiepolitik; Tiroler Landesregierung, Abteilung Wasser-, Forst- und Energierecht: Innsbruck, Austria, 2008.
35. Bayerisches Staatsministerium für Wirtschaft, Infrastruktur, Verkehr und Technologie. Eckpunkte der bayerischen Energiepolitik; Bayerisches Staatsministerium für Wirtschaft, Infrastruktur, Verkehr und Technologie: Munich, Germany, 2008.
36. Strasser, U.; Mauser, W. Modelling the spatial and temporal variations of the water balance for the Weser catchment 1965–1994. J. Hydrol. 2001, 254, 199–214.
37. Prasch, M.; Mauser, W. GLOWA-Danube: Integrative Techniques, Scenarios and Strategies Fort he Future of Water in the Upper Danube Basin. In Proceedings of the XXVth Conference of the Danubian Countries in Budapest, Budapest, Hungary, 16–17 June 2011.
38. Prasch, M.; Marke, T.; Strasser, U.; Mauser, W. Large scale integrated hydrological modelling of the impact of climate change on the water balance with DANUBIA, Adv. Sci. Res. 2011, 7, 61–70.
39. Prasch, M. Distributed Process Oriented Modelling of the Future Impact of Glacier Melt Water on Runoff in the Lhasa River Basin in Tibet. Ph.D. Thesis, Faculty of Geosciences, LMU Munich, Munich, Germany, 2010. Available online: http://edoc.ub.uni-muenchen.de/13031/ (accessed on 25 August 2011).
40. Strasser, U.; Franz, H.; Mauser, W. Distributed Modelling of Snow Processes in the Berchtesgaden National Park (Germany). In Proceedings of the Alpine Snow Workshop, Munich, Germany, 5–6 October 2006; Volume 53, pp. 117–130.

41. Bayerisches Landesamt für Umwelt. Québec-Bavarian International Collaboration on Climate Change—Klimawandel und Flussgebietsmanagement, Hof, Germany, 2011. Available online: http://www.lfu.bayern.de/wasser/klima_wandel/projekte/kliflum/index.htm (accessed on 25 August 2011).
42. Climb Project. Climate Induced Changes on the Hydrology of Mediterranean Basins. Reducing Uncertainty and Quantifying Risk through an Integrated Monitoring and Modelling System, 2010. Available online: http://www.climb-fp7.eu (accessed on 25 August 2011).
43. Blöschl, G.; Montanari, A. Climate change impacts—Throwing the dice? Hydrol. Process. 2010, 24, 374–381.
44. Beven, K. Changing ideas in hydrology—The case of physically-based models original research article. J. Hydrol. 1989, 105, 157–172.
45. Cunge, J.A. On the subject of a flood propagation computation method (Muskingum method). J. Hydraul. Res. 1969, 7, 205–230.
46. Todini, E. A mass conservative and water storage consistent variable parameter Muskingum-Cunge approach. Hydrol. Earth Syst. Sci. 2007, 4, 1549–1592.
47. Ostrowski, M.; Lohr, H. Modellgestützte bewirtschaftung von talsperrensystemen Wasser und Abfall 2002, 1–3, 40–45.
48. Nash, J.E.; Sutcliffe, J.V. River flow forecasting through conceptual models, a discussion of principles. J. Hydrol. 1970, 10, 282–290.
49. Schaefli, B.; Gupta, H.V. Do Nash values have value? Hydrol. Process. 2007, 21, 2075–2080.
50. Wenn bei Starkregen die Produktion einbricht. Energiespektrum 2010, 9–10, 40–41. Available online: http://www.energiespektrum.de/index.cfm?pid=1705&pk=98082 (accessed on 27 September 2011).
51. EEX transparency platform publishing market-relevant energy generation and consumption data. European Energy Exchange AG: Leipzig, Germany 2011. Available online: http://www. transparency.eex.com/en/ (accessed on 28 June 2011).
52. Mauser, W.; Prasch, M.; Strasser, U. Physically based Modelling of Climate Change Impact of Snow Cover Dynamics in Alpine Regions Using a Stochastic Weather Generator. In Proceedings of the International Congress on Modelling and Simulation MODSIM07, Christchurch, New Zealand, 10–13 December 2007; pp. 2138–2145.
53. Orlowsky, B.; Gerstengarbe, F.W.; Werner, P.C. A resampling scheme for regional climate simulations and its performance compared to a dynamic RCM. Theor. Appl. Climatol. 2007, 92, 209–223.
54. Yates, D.; Gangopadhyay, S.; Rajagopalan, B.; Strzepek, K. A technique for generating regional climate scenarios using a nearest neighbour algorithm. Water Resour. Res. 2003, 39, 1–15.
55. Buishand, T.A.; Brandsma, T. Multiscale simulation of daily precipitation and temperature in the Rhine Basin by nearest-neighbour resampling. Water Resour. Res. 2001, 37, 2761–2776.
56. Young, K.C. A multivariate chain model for simulating climatic parameters from daily data. J. Appl. Meteorol. 1994, 33, 661–671.
57. Richardson, C.W. Stochastic simulation of daily precipitation, temperature and solar radiation. Water Resour. Res. 1981, 17, 182–190.

58. Semenov, M.A.; Brooks, R.J.; Barrow, E.M.; Richardson, C.W. Comparison of WGEN and LARS-WG stochastic weather generators for diverse climates. Clim. Res. 1998, 10, 95–107.
59. Racsko, P.; Szeidl, L.; Semenov, M.A. A serial approach to local stochastic weather models. Ecol. Model. 1991, 57, 27–41.
60. Jacob, D.; Goettel, H.; Kotlarski, S.; Lorenz, P.; Sieck, K. Klimaauswirkungen und Anpassungen in Deutschland. In Climate Change; Federal Environmental Agency: Dessau-Roßlau, Germany, 2008; Volume 11/08.
61. Frei, C.; Christensen, J.H.; Déqué, M.; Jacob, D.; Jones, R.G.; Vidale, P.L. Daily precipitationstatistics in regional climate models: Evaluation and intercomparison for the European Alps. J. Geophys. Res. 2003, 108, doi:10.1029/2002JD002287.
62. Beniston, M.; Stephenson, D.B.; Christensen, O.B.; Ferro, C.A.T.; Frei, C.; Goyette, S.; Halsnaes, K.; Holt, T.; Jylhä, K.; Koffi, B.; et al. Future extreme events in European climate: An exploration of regional climate model projections. Clim. Change 2007, 81, 71–95.
63. ENSEMBLES: Climate Change and Its Impacts Summary of Research and Results from the ENSEMBLES Project; van der Linden, P., Mitchell, J.F.B., Eds.; Met Office Hadley Centre: Exter, UK, 2009; p. 164. Available online: http://www.ensembles-eu.org (accessed on 25 August 2011).
64. Jacob, D.; Bärring, L.; Christensen, O.B.; Christensen, J.H.; de Castro, M.; Déqué, M.; Giorgi, F.; Hagemann, S.; Hirschi, M.; Jones, R.; et al. An inter-comparison of regional climate models for Europe: model performance in present-day climate. Clim. Change 2007, 81, 31–52.
65. Pfeiffer, A.; Zaengl, G. Validation of climate-mode MM5-simulations for the European Alpie region. Theor. Appl. Climatol. 2009, 101, 93–108.
66. Reiter, A.; Weidinger, R.; Mauser, W. Recent climate change at the Upper Danube—A temporal and spatial analysis of temperature and precipitation time series. Clim. Change 2011, doi:10.1007/s10584-011-0173-y.
67. Mann, H.B. Nonparametric test against trends. Econometrica 1945, 13, 245–259.
68. Kendall, M.G. Rank Correlation Methods, 4th ed.; Griffin: London, UK, 1970.
69. Mauser, W.; Marke, T. Climate Change and Water Resources: Scenarios of Low-flow Conditions in the Upper Danube River Basin. IOP Conf. Ser. 2008, 4, doi:10.1088/1755-1307/4/1/012027.
70. Die Wettbewerbsfähigkeit von großen Laufwasserkraftwerken im liberalisierten deutschen Strommarkt. Endbericht für das Bundesministerium für Wirtschaft und Arbeit (BMWA); Fichtner GmbH & Co KG: Stuttgart, Germany, 2003. Available online: http://www.emissionshandelfichtner.de/pdf/BMWA_Langfassung.pdf (accessed on 12 May 2011).
71. Mittlere Donaukraftwerke AG (MDK); BEW (Bayerische Elektrizitätswerke GmbH): Augsburg, Germany, 2011. Available online: http://www.bew-augsburg.de/cms_bew_inter/kraftwerke/wasserkraft/technikmdk.asp (accessed on 12 May 2011).
72. Strasser, K.-H. Aktuelle Neubauprojekte der E.ON Wasserkraft—Verändertes Umfeld und deren Auswirkungen. Wasserwirtschaft 2007, 10, 102–104.
73. Das Kraftwerk Kaunertal; TIWAG (Tiroler Wasserkraft AG): Innsbruck, Austria, 2006.

74. Wasserkraftausbau; TIWAG (Tiroler Wasserkraft AG): Innsbruck, Austria, 2011. Available online: http://www.tiroler-wasserkraft.at/de/hn/wasserkraftausbau/index.php (accessed on 9 June 2011).
75. E.ON Wasserkraft GmbH and Bayerische Elektrizitätswerke GmbH. Masterplan. Ausbaupotentiale Wasserkraft in Bayern. Bericht aus Sicht der beiden großen Betreiber von Wasserkraftanlagen in Bayern; Landshut: Augsburg, Germany, 2009. Available online: http://www.lfu.bayern.de/wasser/fachinformationen/fliessgewaesser_wasserkraft anlagenstatistik/doc/potentialstudie.pdf (accessed on 9 June 2011).

PART IV

GEOTHERMAL ENERGY AND CLIMATE CHANGE

CHAPTER 8

Geothermal Power Growth 1995–2013: A Comparison with Other Renewables

LADISLAUS RYBACH

8.1 INTRODUCTION

Renewable sources of electricity are generating increasing interest and having a corresponding impact on the energy scene. Geothermal energy sources have the advantage of providing base-load electricity, i.e., independent of daily, seasonal or climatic variations and thus can complement other, intermittently producing renewable sources like wind or solar. Whereas wind and solar energy sources are abundant on the surface, for geothermal sources one has to go deep, usually a few kilometers. In the following, geothermal electricity is addressed and the global electricity supply from various renewable sources will be presented and compared. The growth rate in renewable power generation is a decisive factor on the electricity market.

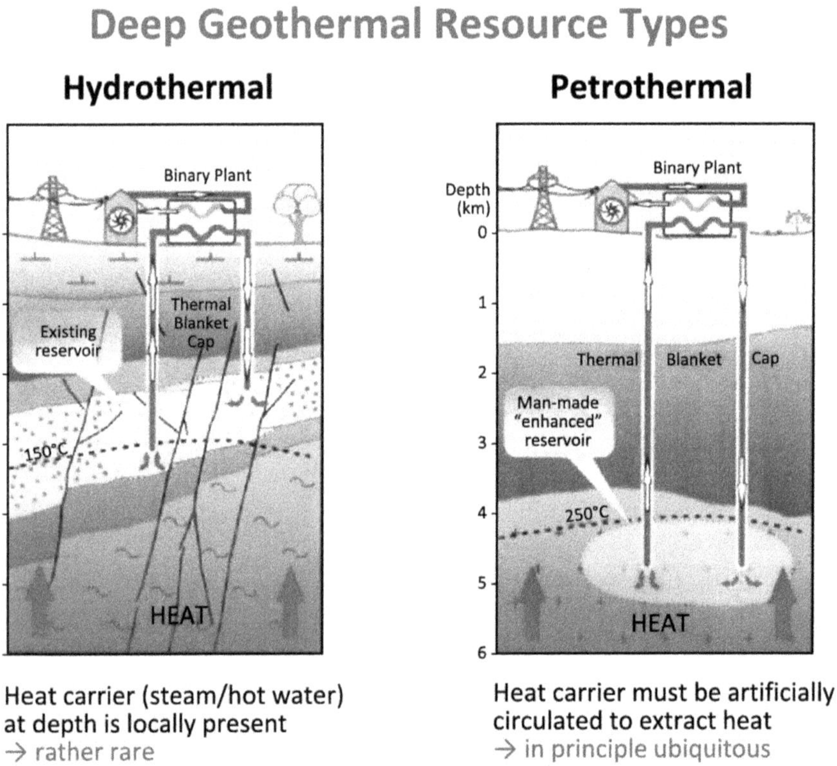

FIGURE 1: The two types of geothermal resources capable to generate electric power. The hydrothermal type on a natural reservoir that can feed, besides binary power plants, also geothermal steam condensing turbines. For the petrothermal type the reservoir needs to be created (details see text). Modified from Figure 3 in Geothermal Electricity (GEOELEC) Resource Assessment Protocol [1].

8.2 GEOTHERMAL POWER GENERATION

There are two main types of deep geothermal resources from which electricity can be produced: hydrothermal and petrothermal. Hydrothermal resources have naturally occurring geothermal fluids at depth, often originating from surface infiltration of precipitation. The fluids can be used as heat carriers and taken out from the ground through boreholes. Such hydrothermal resources like deep aquifers exist only when specific geologic/hydrogeologic conditions prevail, which makes them rather rare. Petrothermal resources on the other hand are more or less ubiquitous and immense; they consist basically of the "heat in place" in deep rock formations. The heat must be therefore extracted, e.g., by establishing a fluid circulation through a special, man-made heat exchanger at depth (see below for details). So far 99.99% of all existing geothermal power plants use hydrothermal resources. Figure 1 shows schematically the two resource types.

Geothermal power plants provide base-load electricity. Currently, the total globally installed capacity amounts to about 12 GWe, in 24 countries, with a total production of 76 TW·h/yr [2]. So far, practically all power plants use hydrothermal resources. Geothermal power generation started in 1904 in Larderello, Italy. In earlier days, reservoirs with dry steam have been tapped, later also those with steam/water mixtures. Such high-temperature fields (>200 °C in less than 2 km depth) are mostly located in volcanic areas and are correspondingly rare. The average power plant size is about 50 MWe. The largest hydrothermal plant to date, at Toanga (previously called Nga Awa Purua) in New Zealand operates with a single 140 MWe turbine unit and is fed by only six production wells [3].

With advanced technology such as binary power plants it is now possible to convert heat to power also with lower fluid temperatures (100–120 °C). But the conversion efficiency is correspondingly low (a few percentage points only) and the plant size is also limited (only a few MWe).

Below a global comparison is presented between geothermal and the other renewable energies, in terms of both potential and power generation. Development growth is presented for wind, solar photovoltaic (PV) and geothermal power and compared for the time period 1995–2013. In addition, a comparison is made of the annual geothermal production in 2013 with the renewables hydropower, biomass, solar PV and wind.

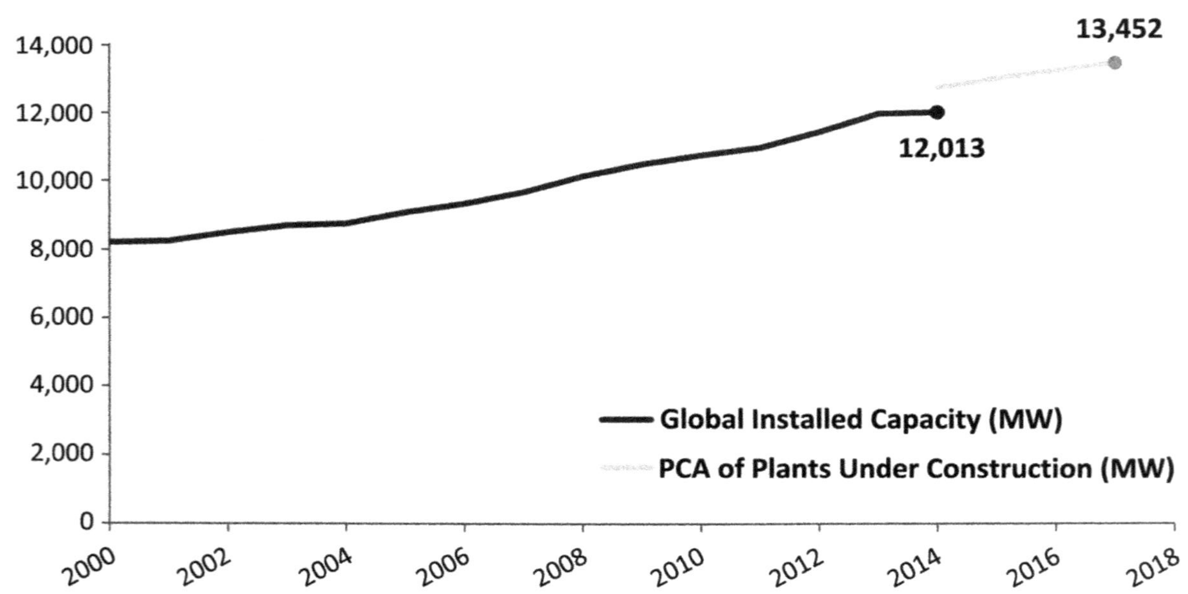

FIGURE 2: Growth of installed geothermal power (MWe) worldwide over the years 2000–2014 (from Geothermal Energy Association (GEA) 2014 [6]). Global growth 2004–2012 ~4%.

FIGURE 3: Growth in global wind power development (GWe) over the years 2000–2013 from REN21 [2].

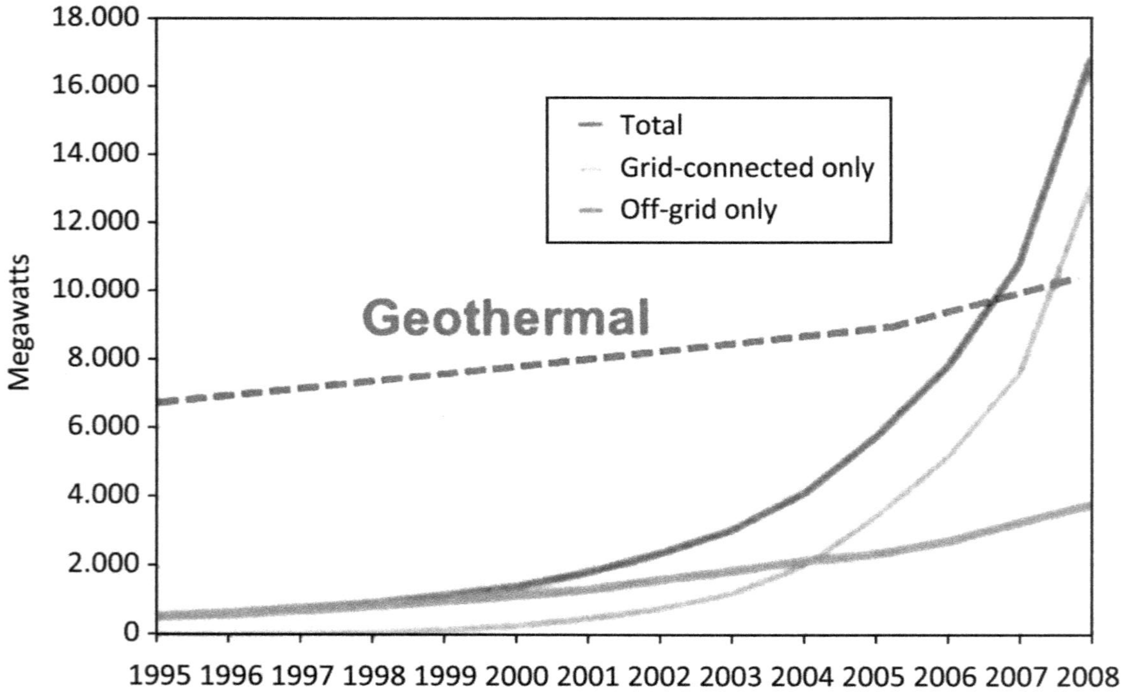

FIGURE 4. Global growth in solar photovoltaic (PV) development (MWe) 1995–2008 from REN21 2009 [7]. Geothermal power growth is plotted for comparison (data from [5]). Until about 2007 geothermal power was far ahead of solar PV.

8.3 LARGE GEOTHERMAL POTENTIAL

A highly respected source (World Energy Assessment (WEA)—a collaborative effort between United Nations Development Programme (UNDP), United Nations Development of Economic and Social Affairs (UN DESA) and the World Energy Council (WEC)) attests the largest potential value to geothermal energy among all forms of renewable energy sources. The comparison is given in Table 1.

TABLE 1: Potential of renewable energy sources, from World Energy Assessment (WEA) [4].

Energy source	Capacity (EJ/yr)
Geothermal	5000
Solar	1575
Wind	640
Biomass	276
Hydro	50
Total	7541

The values are given in capacity units, i.e., energy per unit time. It is obvious that geothermal energy has the largest capacity, although the accuracy of the reported number is limited. This potential is so far only marginally developed.

8.4 GROWTH COMPARISON OVER THE TIME PERIOD 1995–2013

Geothermal power development data is available for the time period 1960–2013 according to Geothermal Energy Association (GEA) 2012 [5] and 2000–2013 from GEA 2014 [6]. The growth is practically linear, with only small increase rate changes lately, see Figure 2.

TABLE 2: Comparison of global electricity production by renewable technologies in 2013 (data from REN21 [2]).

Technology	Installed capacity		Annual production		Availability
	GWe	%	TW h/yr	%	%
Hydropower	1000	64.2	3680	74.9	42
Biomass	88	5.7	405	8.2	53
Wind	318	20.4	585	11.9	21
Geothermal	12	0.8	76	1.5	72
Solar PV	139	8.9	170	3.5	14
Total	1557	100	4916	100	-

New data on the development of power generation from renewable sources is given in REN21 2014 [2]. The installed capacity of wind power shows a clearly accelerating trend of an exponential nature (Figure 3), with an annual growth rate of about 25%.

A similar trend of exponential growth is reported for solar PV power, both grid-connected and off-grid production already for 1995–2008. In Figure 4, the geothermal power growth in the same period—from [5]—is plotted for comparison. It is evident that geothermal had the lead over solar PV in the time before year 2007. Afterwards solar PV clearly took over.

Now new solar PV data are available. Figure 5 shows the situation by end of 2013. For comparison, the geothermal data are plotted again. It is clear that geothermal is now left far behind. Here it must be noted that practically all geothermal power originates so far from hydrothermal resources.

Here it must be emphasized that the Figure 2, Figure 3, Figure 4 and Figure 5 refer to installed capacity, not to actual power production. What counts is the produced amount of electricity. Annual production data (in TW·h) are assembled in Table 2 for various renewable sources. Wind is not blowing at all times; the sun is shining only during daytime whereas geothermal production can go on at practically all times (except for production stops, for example during maintenance operations). This is reflected by the capacity factor (basically the percentage of yearly operating hours), given in Table 2. Sometime in 2011 solar PV took also over geothermal electricity in terms of global annual production; and the solar-geothermal gap is thus further increasing.

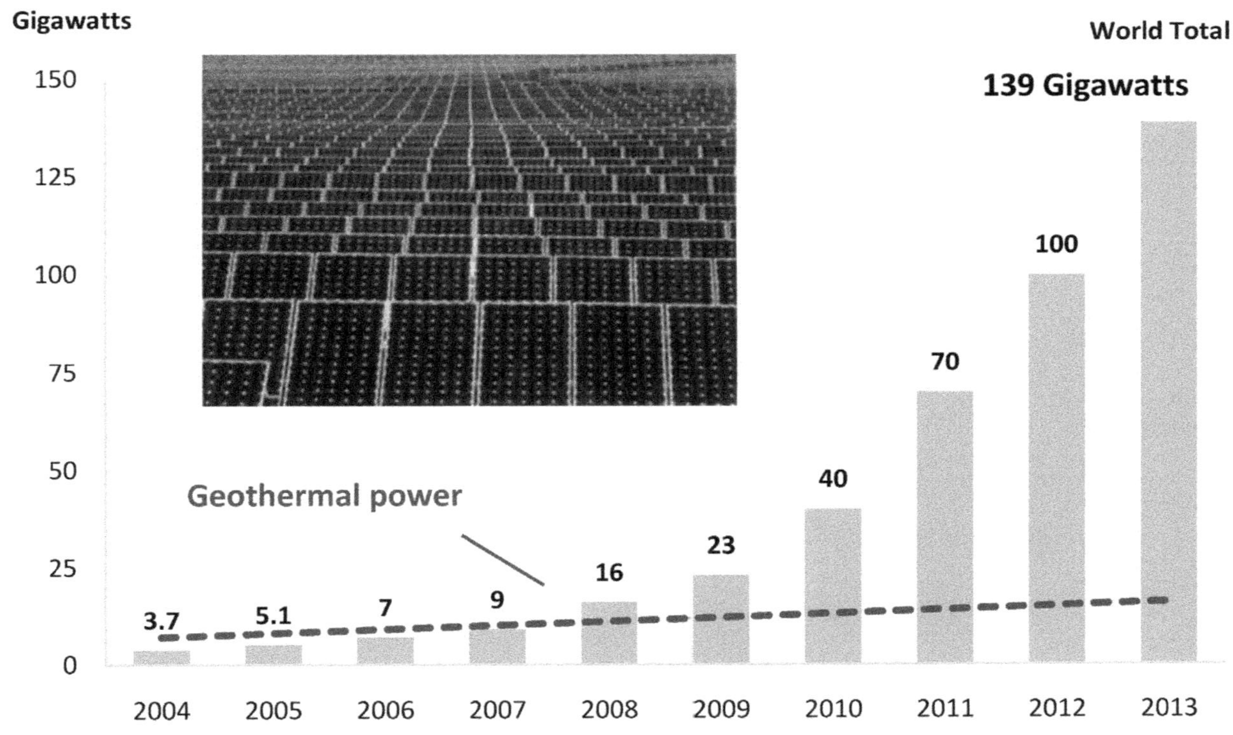

FIGURE 5: Global growth in solar PV development (MWe) until 2013 from REN21 2014 [2]. Geothermal power growth (dashed line) is plotted for comparison—from Figure 2. Whereas solar growth is annually about 40%, geothermal growth remains low at 4% per year.

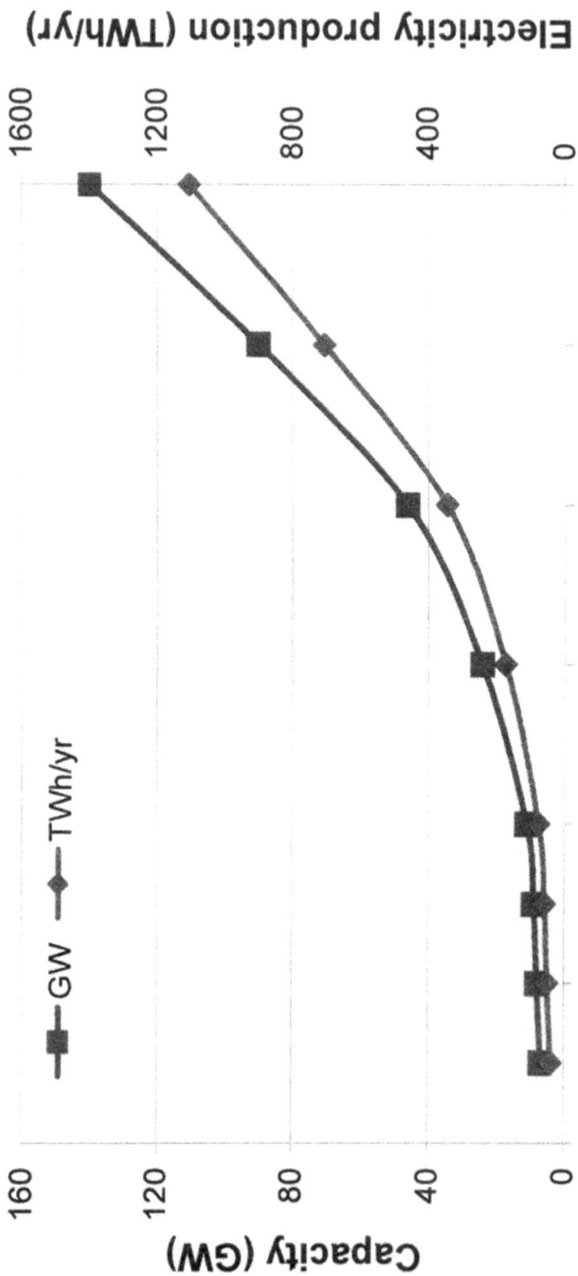

FIGURE 6: Installed geothermal capacity and electricity production since 1995 and forecasts for 2010–2050. From [8].

From the above comparison it is evident that currently geothermal power development is left behind wind and solar PV. Whereas geothermal development growth is more or less linear (steady but slow growth—just a few percent increase per year), wind and solar PV exhibit accelerating growth with a clearly exponential tendency. To keep pace geothermal growth needs to be speeded up too; in the following some possible ways and means to accomplish this are addressed, primarily for power generation.

8.5 HOW TO ACHIEVE ACCELERATED GEOTHERMAL POWER GROWTH?

Until today the growth in installed geothermal power capacity originated entirely from "conventional", hydrothermal resources. Such resources are found in numerous but special places, with high-temperature geothermal fluids present in the subsurface at relatively shallow depths (2–4 km). Such "anomalous" places can mainly be found in volcanic terranes or in other regions, depending on their plate tectonic settings (details see e.g., in [8]). It can be expected that geothermal power development based on conventional high-enthalpy resources will remain more or less linear in the future; therefore some new technology is needed to provide the exponential growth component. In the following the case is made that enhanced geothermal systems (EGS) technology could play this role.

In a study commissioned by the Intergovernmental Panel on Climate Change (IPCC) a team of authors [8] estimated the growth curve in geothermal power development from the present to year 2050. Figure 6 shows the result (installed capacity as well as power production). The curves in Figure 6 also exhibit exponential character.

8.6 EGS TECHNOLOGY: GOALS AND OPEN QUESTIONS

The renowned Massachusetts Institute of Technology (MIT) [9] study "The Future of Geothermal Energy—Impact of EGS on the United States in the 21th Century" suggests that EGS will be the future of geothermal energy utilization. EGS is an umbrella term for various other denotations such as Hot Dry Rock, Hot Wet Rock, and Hot Fractured Rock. The MIT study determined EGS resources >200,000 EJ alone for the USA, corresponding to 2000 times the annual primary energy demand.

The EGS principle is simple: in the deep subsurface where temperatures are high enough for power generation (150–200 °C) an extended, well distributed fracture network is created and/or enlarged to act as new fluid pathways and at the same time as a heat exchanger ("reservoir"). Water from the surface is pumped through this deep reservoir using injection wells and recovered by production wells as steam/hot water. The extracted heat can be used for district heating and/or for power generation.

The core piece of an EGS installation is the heat exchanger at depth. It is generally accepted that it must have a number of properties in order to be technically feasible and economically viable. These refer to the total volume, the total heat exchange surface, the flow impedance, and the thermal and stress-field properties. The key properties are summarized in Table 3.

TABLE 3. Required properties for an enhanced geothermal systems (EGS) reservoir (after RHC Platform [10]).

Fluid production rate	50–100 kg/s
Fluid temperature at wellhead	150–200 °C
Total effective heat exchange surface	$>2 \times 106\ m^2$
Rock volume	$>2 \times 108\ m^3$
Flow impedance	<0.1 MPa/(kg/s)
Water loss	<10%

Although the minimum requirements for an economically viable EGS reservoir are herewith set, their realization in a custom-made manner to comply with differing site conditions is not yet demonstrated. The key issue is the development of a technology to produce electricity and/or heat from a basically ubiquitous resource, in a manner relatively independent of local subsurface conditions, i.e., to develop a technology for the creation of EGS downhole heat exchangers—wherever needed—with the properties quantified above. Therefore, several questions about establishing and operating EGS heat exchangers that are still open need to be addressed and answered. Here are some of the key issues:

- Development of a technology to produce electricity and/or heat from a basically ubiquitous resource, in a manner more or less independent of site conditions.
- Site exploration must clarify the local temperature and stress field, lithology, kind and degree of already existing fracturation, natural seismicity.
- In creating EGS heat exchangers at several kilometers depth, questions of rock mechanics like the role of anisotropy degree, stress change propagation/transmission—fast/"dry"? slow/"wet"? (under different site conditions)—need to be answered.
- EGS induced seismicity (during stimulation in establishing the EGS heat exchanger but also during production) becomes a real issue, and thus needs to be controlled. Magnitudes need to be limited since public acceptance will be decisive [11].
- Uniform connectivity throughout a planned reservoir cannot yet be engineered. There is no experience with possible changes of an EGS heat exchanger over time; permeability enhancement (e.g., new fractures generated by cooling cracks) could increase the recovery factor while permeability reduction (e.g., by mineral reactions) or short-circuiting could reduce recovery.
- This leads to the question of production sustainability. The production level needs to be set in order to guarantee longevity of the system (details in [12]).

8.7 INCREASING EGS POWER PLANT SIZE

In order to play a significant role on the electricity scene, geothermal power plants should have the size of at least some tens of MWe. So far, EGS plants (Soultz sous Forêts in France, Landau and Insheim in Germany) have just a few MWe installed capacity. Wind generators nowadays come with at least 2 MWe and can be installed, especially offshore, in large numbers.

One of the main future R&D goals will be to work out how and to what extent could the EGS power plant size be upscaled. So far, there are only some theoretical calculations available; see e.g., [13]. In this publication an EGS scheme with 24 injection and 19 production wells is modelled, providing a net power output of around 60 MWe. Of course such ideas need to become substantiated by field evidence.

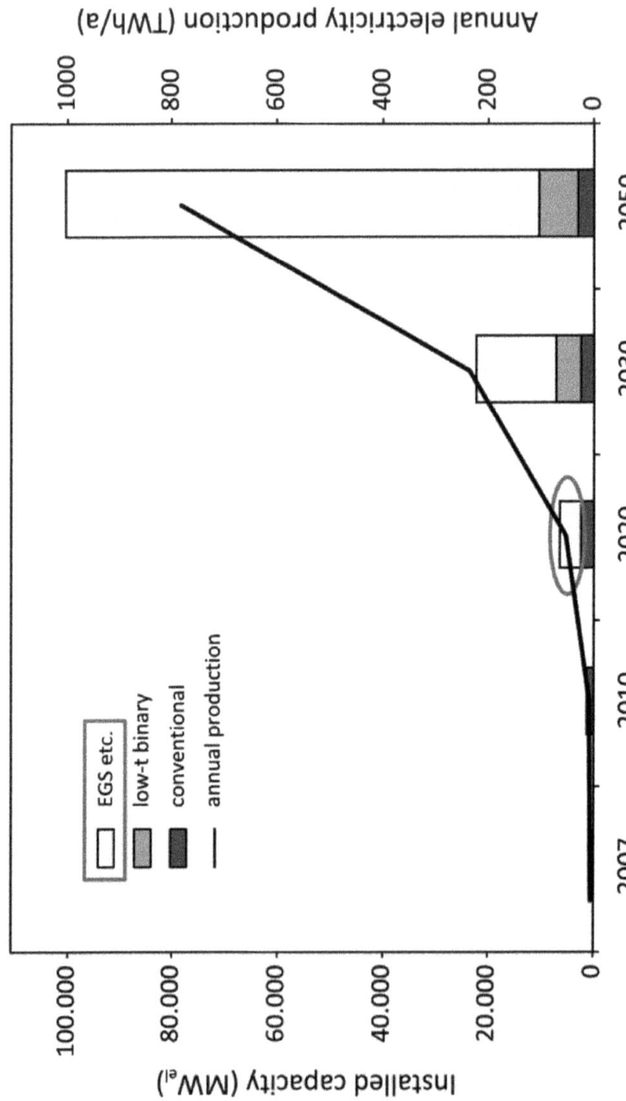

FIGURE 7: Vision of European Geothermal Energy Council (EGEC) about geothermal electricity growth until 2050, from EGEC 2012 [14]. The largest share should come from EGS.

8.8 EUROPEAN GEOTHERMAL GROWTH PERSPECTIVES, FINANCIAL ASPECTS

The European Union has ambitious goals in term of renewable energy growth; the "20-20-20 goal" (20% share of renewable energies, 20% energy savings and 20% CO_2 emission reduction until 2020) clearly calls also for more geothermal electricity. In the Union, a goal of 3 GWe EGS capacity has been proposed for the year 2020 and further substantial EGS growth by European Geothermal Energy Council (EGEC) [14], see Figure 7.

What financial sources would be needed for the realization of this vision? Currently the cost of establishing a generating capacity of 1 MWe from EGS (including exploration, drilling, stimulation, power plant, etc.) is estimated to be around 16 million € according to Geothermal Electricity (GEOELEC) [15]. Thus the 3 GWe EGS capacity foreseen for 2020 in Europe (circled in Figure 7) would require an investment of about 50 billion €. Today it is unclear where such a funding would come from.

It is obvious from the above-described knowledge gaps that very substantial R&D efforts are still needed to make EGS become the future of geothermal energy. Whereas some of the problems could be tackled by broad-based international cooperation, national R&D programs have to provide additional means for the challenge. Public funding, mainly by governmental agencies, will be indispensable.

Although envisaged for conventional geothermal resources, Ibrahim [16] describes five steps to expedite development; four of them are based on fund allocations from national or regional governments. In any case it will be crucial to make rapid progress in tackling and solving the above-mentioned, still open financing problems.

8.9 CONCLUSIONS AND OUTLOOK

Geothermal power develops steadily world-wide, albeit with modest growth rates (a few % per year). In some countries, like in Iceland, New Zealand or Turkey, the growth is remarkable. At the same time, wind and solar PV develop exponentially, with 30%–40% annual growth. In other words: globally, geothermal power falls increasingly back behind electricity from wind and solar PV.

Therefore, geothermal power growth should be accelerated. Since the development of hydrothermal resources cannot be accelerated much, mainly because such resources are limited; the only option that remains are petrothermal resources. The only problem: how to get out the heat in place? In particular, the following questions need to be addressed:

- Where? (favorable site conditions → exploration)
- How? (sufficient, deep heat exchanger realization → proper, site dependent stimulation—without significant induced seismicity)
- With what efficiency? (recovery factor → enhancement of heat extraction, production sustainability). Recovery factor, R (%) = extractable heat/heat in place

Besides, upscaling EGS power plant size will be decisive. EGS pilot plants are badly needed, as is long-term experience. In addition, the financing of all the R&D needs should also get solved. All these open questions need to be answered—and rather quickly so.

The future of geothermal power will strongly depend on to what extent can be the power plant deployment accelerated. Other sources of renewable energy are developing rapidly, especially wind and solar PV: wind energy recently accomplished to install 35 GWe additional capacity per year; solar PV reached 39 GWe/yr, whereas geothermal power growth remains below 2 GWe/yr. Even when one takes into account the higher geothermal capacity factor the need for speeding-up geothermal development is obvious. Accelerating EGS development could provide a break-through, under the condition that the necessary significant funding needs can be met. This, in turn, will require heavy engagement of both the public and the private sector.

REFERENCES

1. Van Wees, J.D.; Calcagno, P.; Dezayes, C.; Lacasse, C. A Methodology for Resource Assessment and Applications to Core Countries; Geothermal Electricity (GEO-ELEC): Brussels, Belgium, 2013; Available online: http://www.geoelec.eu/concep/library/(accessed on 22 July 2014).

2. Renewables 2014: Global Status Report 2014; Renewable Energy Policy Network for the 21st Century (REN21): Paris, France, 2014.
3. Tadao, H.; Muto, T. Technical Features of Nga Awa Purua Geothermal Power Station New Zealand. In Proceedings of Renewable Energy 2010, Yokohama, Japan, 27 June–2 July 2010; Abstract O-Ge-4-4. Available online: http://www.re2010.org/ (accessed on 22 July 2014).
4. World Energy Assessment (WEA) Report: Energy and the Challenge of Sustainability; United Nations Development Programme (UNDP), Bureau for Development Policy, One United Nations Plaza: New York, NY, USA, 2000.
5. Geothermal: International Market Overview Report; Geothermal Energy Association (GEA): Washington, DC, USA, 2012.
6. Annual U.S. & Global Geothermal Power Production Report; Geothermal Energy Association (GEA): Washington, DC, USA, 2014.
7. Renewables 2009: Global Status Report 2009; Renewable Energy Policy Network for the 21st Century (REN21): Paris, France, 2009.
8. Fridleifsson, I.B.; Bertani, R.; Huenges, E.; Lund, J.W.; Ragnarsson, A.; Rybach, L. The Possible Role and Contribution of Geothermal Energy to the Mitigation of Climate Change. In Proceedings of the IPCC Scoping Meeting on Renewable Energy Sources, Luebeck, Germany, 20–25 January 2008; Volume 20, pp. 59–80.
9. Tester, J.W.; Anderson, B.J.; Batchelor, A.S.; Blackwell, D.D.; DiPippo, R.; Drake, E.M.; Garnish, J.; Livesay, B.; Moore, M.C.; Nichols, K.; et al. The Future of Geothermal Energy—Impact of Enhanced Geothermal Systems (EGS) on the United States in the 21st Century; Massachusetts Institute of Technology (MIT): Cambridge, MA, USA, 2006; p. 358. Available online: http://www1.eere.energy.gov/library/default.aspx?page=4 (accessed on 22 July 2014).
10. Strategic Research Priorities for Geothermal Technology; European Technology Platform on Renewable Heating and Cooling (RHC): Brussels, Belgium, 2012; p. 65. Available online: http://www.rhc-platform.org/publications/ (accessed on 22 July 2014).
11. Majer, E.; Baria, R.; Stark, M. Protocol for Induced Seismicity Associated with Enhanced Geothermal Systems; Report Produced in Task D Annex I. International Energy Agency—Geothermal Implementing Agreement: Taupo, New Zealand, 2008.
12. Rybach, L.; Mongillo, M. Geothermal sustainability—A review with identified research needs. GRC Trans. 2006, 30, 1083–1090.
13. Vörös, R.; Weidler, R.; de Graaf, L.; Wyborn, D. Thermal Modelling of Long Term Circulation of Multi-Well Development at the Cooper Basin Hot Fractured Rock (HFR) Project and Current Proposed Scale-Up Program. In Proceedings of the Thirty-Second Workshop on Geothermal Reservoir Engineering, Stanford, CA, USA, 22–24 January 2007.
14. European Geothermal Energy Council (EGEC). EGEC Vision for 2050 on Geothermal Power in Europe (Vision of ETP GEOELEC). 2012. Available online: http://egec.info/policy/tp-geoelec/ (accessed on 22 July 2014).
15. Geothermal Investment Guide; Geothermal Electricity (GEOELEC): Brussels, Belgium, 2011; p. 32. Available online: http://www.geoelec.eu/concep/library/ (accessed on 22 July 2014).
16. Ibrahim, H.D. Why is Geothermal Development Slow—How to Accelerate It? Petrominer Monthly Magazine 2009, Volume 7, 28–29.

CHAPTER 9

Modeling of an Air Conditioning System with Geothermal Heat Pump for a Residential Building

SILVIA COCCHI, SONIA CASTELLUCCI, AND ANDREA TUCCI

9.1 INTRODUCTION

The increasing energy demands the fact that fossil fuels are finite resouces and the problem of pollutant emissions has allowed renewable energy sources (RES) to be considered and developed, including geothermal [1].

Geothermal energy is used in order to generate electricity or for direct uses, especially for heating [2]. Direct use of geothermal energy is the oldest and the most common utilization of this energy source.

There are many possibilities for these uses: geothermal heat pumps or ground source heat pumps (GSHPs), space heating, greenhouse and covered ground heating, aquaculture pond and raceway heating, agricultural crops drying, industrial process heat, snow melting and space cooling, bathing, and swimming [3].

The GSHP systems for heating and cooling are considered one of the most energy-efficient and cost-effective renewable energy technology [4]. Moreover these systems allow to reduce the greenhouse gases (GHGs) emissions [5].

Geothermal heat pumps systems use the subsoil (the interior of the earth) as a source and extract heat at low deep [6].

Heat pumps are devices that allow to transfer heat from a lower temperature system to a higher temperature system. Geothermal heat pump or ground source heat pump (GSHP) is a heating and/or cooling system that pumps heat to or from the ground. The GSHP uses earth as a heat source (during winter time) or a heat sink (during summer time). This design takes advantage of the moderate temperatures in the ground to boost efficiency and reduce the operational costs of heating and cooling systems [7]. The major advantage of GSHPs derives from their better performance if compared with traditional system because they takes advantage of a more stable temperature of the heat source during the whole year, thus the coefficient of performance (COP) is increased, and the operational costs of heating and cooling are reduced [8].

The GSHPs systems can be classified as follows:

1. open systems: if groundwater is used as heat transfer fluid,
2. closed systems: if there are some heat exchangers in the underground, and the groundwater is not the heat transfer fluid [9].

Among the closer systems there are many different configurations: horizontal, spiral, and vertical loop [10].

An air conditioning system with geothermal heat pumps consists of three parts [11]:

1. geothermal heat exchangers,
2. heat pump,

3. heat distribution system (radiant floors are particularly suitable because they have the great advantage of working with lower temperature gradients if compared to conventional systems [12]).

The operating mechanism for heating is as follows:

1. the fluid (antifreeze added water), flowing in geothermal heat exchangers, exchanges heat with the ground and comes back heated to surface;
2. fluid transmits its heat to the heat pump and back in the heat exchangers with lower temperature;
3. heat pump transmits its heat to the fluid flowing in radiant floors;
4. radiant floors heat the building.

GSHP must have a heat exchanger in contact with the ground or groundwater in order to extract or dissipate heat. Several major design options are available for these, which are classified according to layout and fluid: direct exchange systems circulate refrigerant underground, closed loop systems use a mixture of antifreeze and water, and open loop systems use natural groundwater. Direct exchange geothermal heat pump is the oldest type of geothermal heat pump technology. In this work a GSHP system with vertical heat exchangers has been studied.

The growth of GSHPs technology was slower than other RES or conventional technologies due to many factors: nonstandardized system designs, significant capital costs if compared with other systems, and limited individuals knowledgeable in the installation of GSHP systems. Nevertheless these problems have been resolved on an ongoing basis, and the acceptance of the GSHP technology is increasing [1].

GSHP systems COP is higher than COP of other heat pump systems: it usually varies from 3 to 6, depending on the earth connection setups, system sized earth characteristics, installation depths, and climate of the area. Instead, for example, for an air source heat pump system the COP range is from 2,3 to 3,5 [1].

GSHP systems have higher initial costs than conventional ones because of the costs of the heat pump unit and the connection with the ground, specially drilling, but the operating costs are lower than conventional systems because of their high efficiency. Therefore the economic feasibility of GSHP systems depends on location, due to the price of electricity, natural gas, and other heating fuel. If these are not expensive, GSHP system may not be the cheapest option. On the contrary when there are low electricity costs, the GSHP systems are economically advantageous [1].

In particular, for example, if electricity is produced by a photovoltaic system the operating costs and the GHG emissions become remarkably low.

Therefore in order to increase the efficiency of the GSHP system and to reduce the costs, the optimal sizing of the plant is needful. The TRNSYS 17 software can be used for the system simulation in order to refine the sizing.

9.2 METHODS

9.2.1 PRESIZING

The studied building is part of a building complex and consists of 14 apartments located over 4 floors.

The building was planned to include a central heating system, for heating and domestic hot water, that uses one heat pump and vertical geothermal heat exchangers. During summer time, the system works in natural cooling operative mode.

The thermal load of the building is 43,2 kW for heating (Table 1). We choose one geothermal heat pump with a power of 47,2 kW, whose fluid consists of water and ethylene glycol (25%), with freezing temperature of −13°C.

Geothermal heat exchangers have been presized according to the procedure described in VDI4640 German law [9]. The procedure of VDI4640 is strictly applied for small sized systems (<30 kW), but we used this procedure for an approximate sizing only. Then the final size has been obtained by the approximate one, by simulation with software TRNSYS 17.

TABLE 1: Data for calculation of thermal load.

Building data		
Thermal zone		D
Number of heating days		167
Daily hours of heating		12
Heated surface (gross)	m^2	3510,61
Outer surface that bounds the volume	m^2	2101,73
Usable area	m^2	930
Climatic data project		
Value of internal project temperature	°C	20

The presizing procedure of VDI4640 is the following [13]:

1. calculation of heating load P_t (= 43.2 kW);
2. definition of temperature of radiant floor (= 35°C);
3. choice of heat pump and definition of the COP (coefficient of performance) with the operating condition B0/W35, COP = 4,33;
4. calculation of the power exchanged with the ground P_{ev} [=COP – 1) P_t/COP = 33.22 kW);
5. extraction from VDI4640 of the specific power extraction P_{ter} (= 30 W/m);
6. calculation of total length of vertical geothermal heat exchangers L (= P_{ev}/P_{ter} = 1107 m);
7. oversized (15%) L = 1273 m;
8. assumption of 12 vertical heat exchangers (each one 100 m long) (Table 2).

9.2.2 SIMULATION

The software TRNSYS 17 has been used in order to simulate the operations system.

TABLE 2: Features of vertical heat exchangers.

Features of vertical exchangers	
Number	12
Length	100 m
Distance	8 m
Type	single U
External diameter U	32 mm
Internal diameter U	29 mm
Bore diameter	12 cm
Distance by centers of U	5 cm

TRNSYS is a complete and extensible simulation environment for the transient simulation of systems, including multizone buildings.

TRNSYS consists of a suite of programs: the TRNSYS Simulation Studio, the Simulation Engine (TRNDll.dll) with its executable (TRNExe.exe), the Building input data visual interface (TRNBuild.exe), and the Editor used to create stand-alone redistributable programs, known as TRNSED applications (TRNEdit.exe) [14].

The main visual interface is the TRNSYS Simulation Studio. Here, we can create projects by drag-and-dropping components to the workspace, connecting them together and setting the global simulation parameters. When you run a simulation, the Studio also creates a TRNSYS input file (a text file, that contains all simulation information but no graphical information).

The Simulation Studio also includes an output manager, by which you control which variables are integrated, printed, and/or plotted, and a log/error manager that allows you to study in detail what happened during the simulation [15].

The system has been represented in TRNSYS Simulation Studio and the Building has been created in TRNBuild.

The Building has been modeled in TRNBuild by dividing it into 16 thermal zones (14 apartments, 1 zone of stairs, and 1 garage) [16].

The simulation with TRNSYS 17 allows the calculation and displays one or more variables on interval of time. Most of variables are studied within annual simulation, but average ground temperature is studied in five-year simulation.

We studied the following variables:

1. thermal loads;
2. air temperature in thermal zones;
3. fluid temperature in input and output from radiant floors;
4. fluid temperature in input and output from geothermal heat exchanger;
5. average ground temperature;
6. COP of heat pump;
7. optimal time step for simulations.

Optimal time step has been studied observing building thermal loads varying with time steps.

9.3 RESULTS

The results of the simulation are represented in Table 3 and in Figure 1.

TABLE 3: Thermal load varying time step simulation.

Time-step	Total thermal load [kWh]
1 hour	68535,845218584532
30 minutes	68677,478548530451
20 minutes	68469,790308356302
10 minutes	65833,167347961783
5 minutes	68533,186713158425
4 minutes	68678,316559261531
3 minutes	68680,138881754820
2 minutes	68676,275556369017
1 minute	68677,942675267512

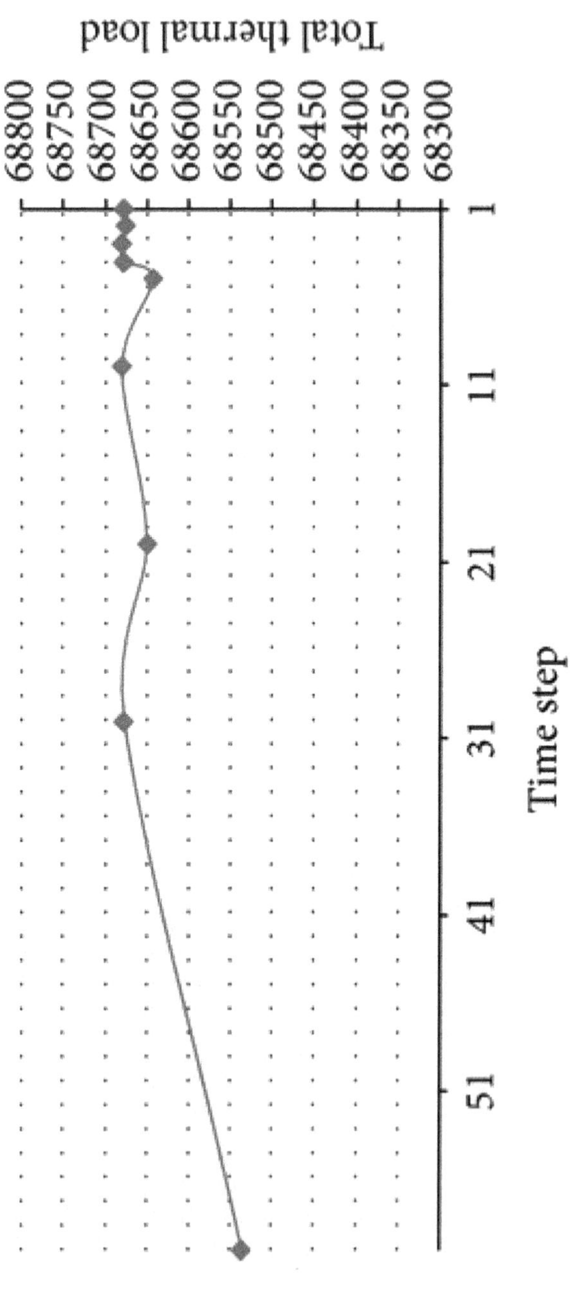

FIGURE 1: Thermal load as a function of time step.

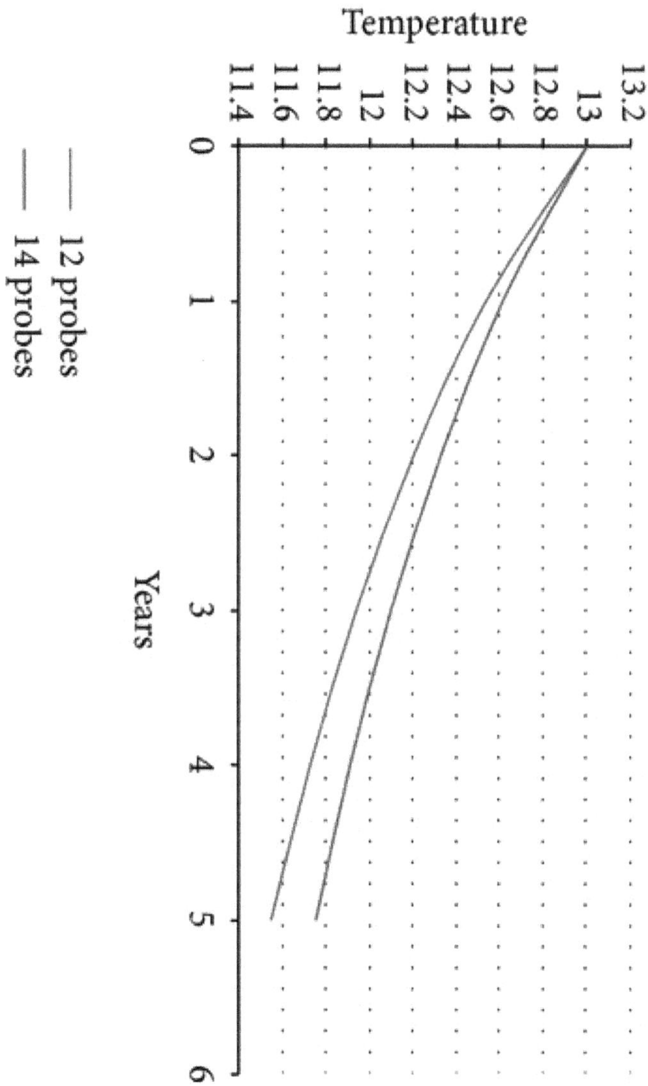

FIGURE 2: Ground temperature variations with exchanger numbers.

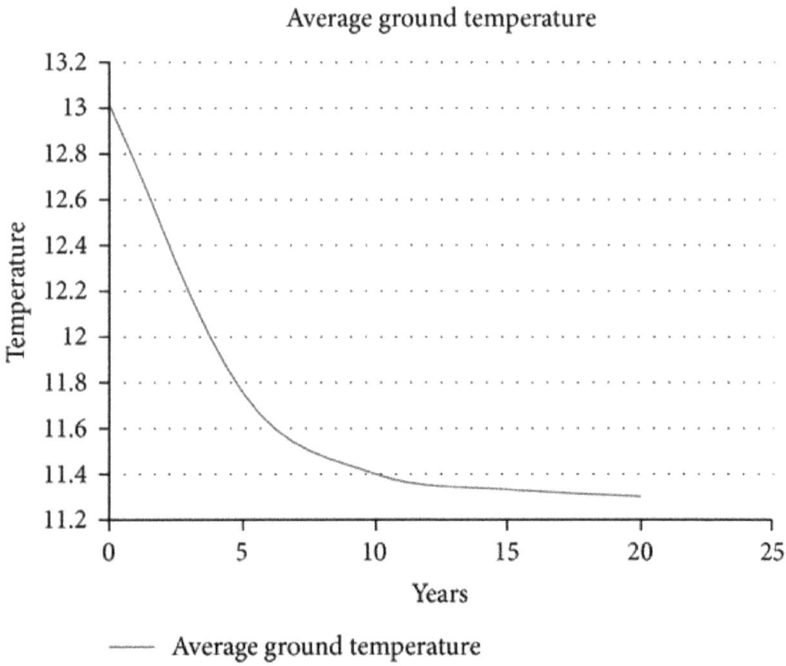

FIGURE 3: Average ground temperature for 14 exchangers in 20 years.

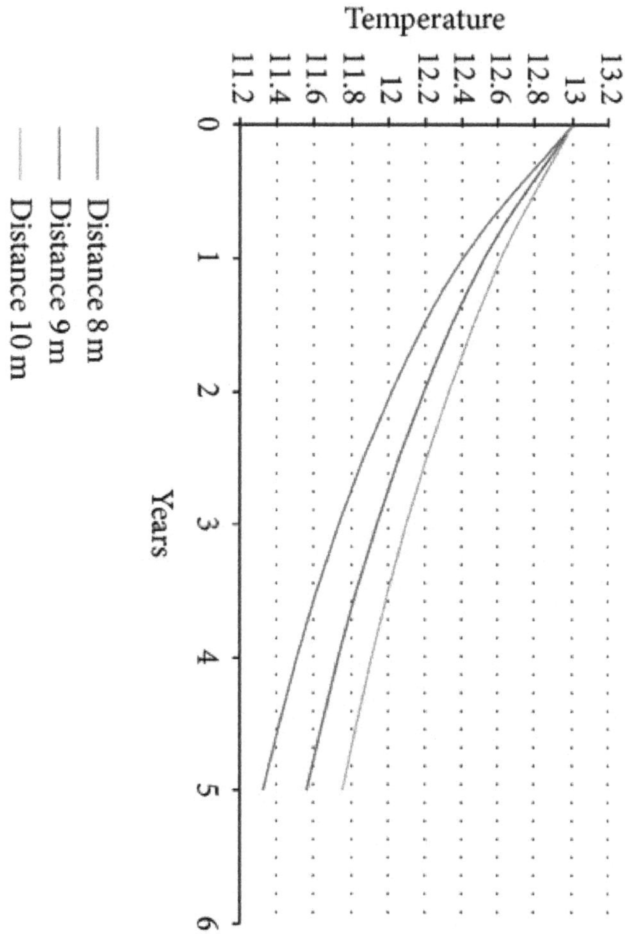

FIGURE 4: Average ground temperature varying number of exchangers.

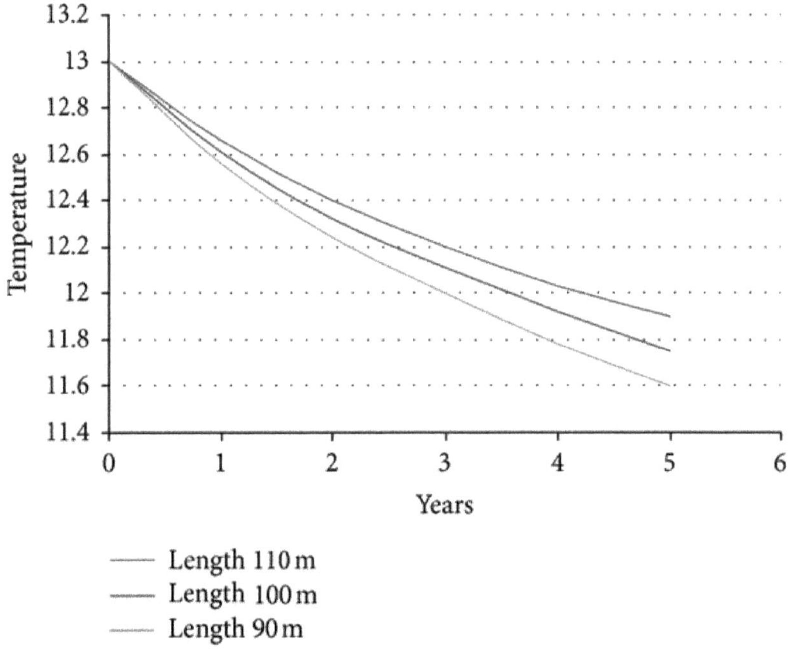

FIGURE 5: Average ground temperature varying exchangers length.

The optimal time step for simulation is 1 minute. The study of average ground temperature is very important because this variable must not change more than 2°C over long period. Simulations are carried out for more conditions (12, 13, and 14 exchangers; distance between exchangers of 8, 9, and 10 m; arrangement of exchangers in series and parallel; length of exchanger of 90, 100, and 110 m).

We have chosen the following as target for the average ground temperature:

1. ensure no overexploitation of the ground;
2. achieve a new balance for the ground average temperature.

Changes in ground temperature varying with number of exchangers are shown in Figure 2.

The average ground temperature change after 20 years is shown in Figure 3, and it is equal to 1,7°C, below the threshold of 2°C, necessary not to have an overexploitation.

Changes in average ground temperature as a function of distance between exchanger are shown in Figure 4.

Changes in average ground temperature varying with exchangers length are shown in Figure 5.

We obtained that the configuration which gives the best performance has the following characteristics: 14 vertical exchangers in series with distance of 10 m and length of 100 m (Table 4).

The temperature of fluid in output from exchangers is represented in Figure 6.

TABLE 4: Configuration of exchangers that gives the best performance.

Number of vertical exchangers	14
Distance	10 m
Length	100 m
Configuration	14 exchangers in series

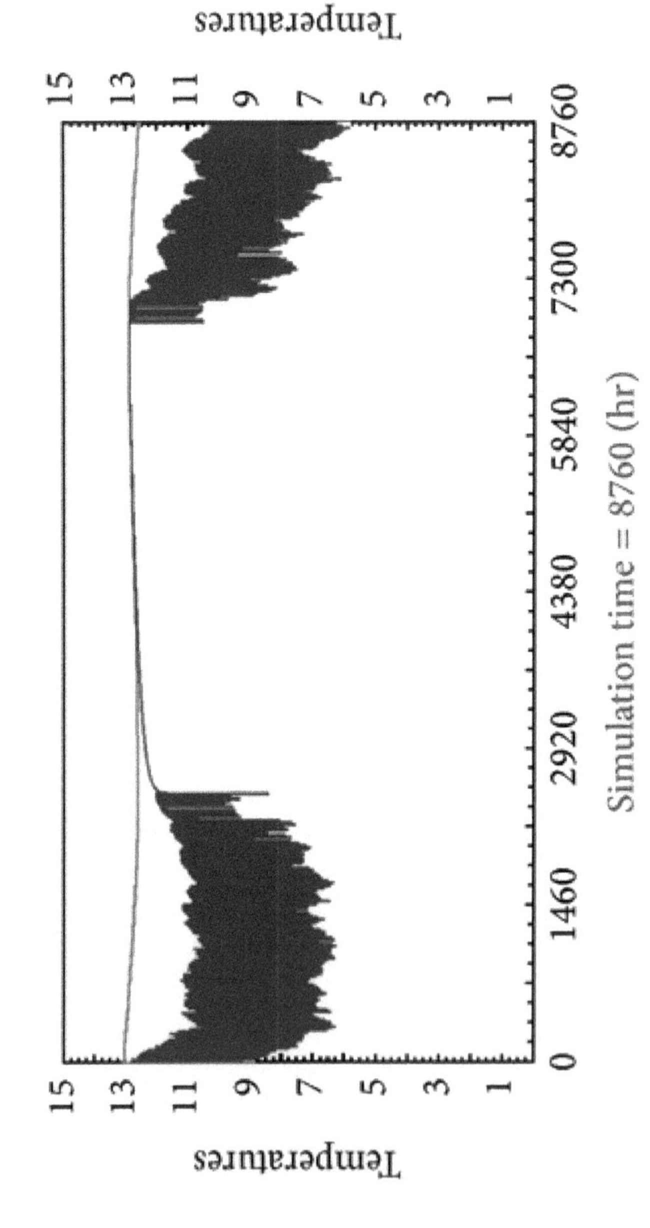

FIGURE 6: Fluid temperature in output from exchangers during the first year.

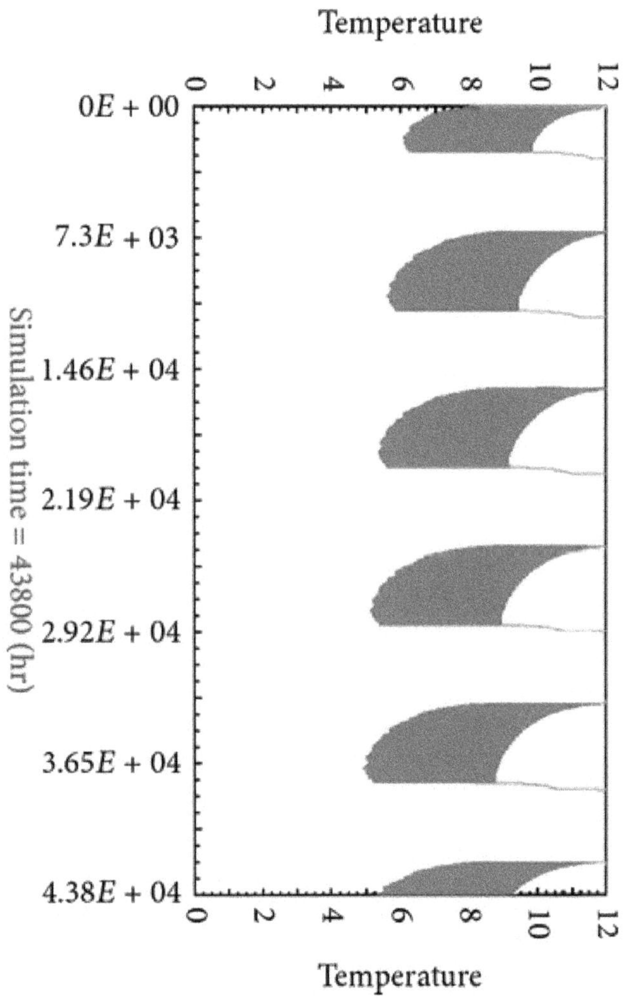

FIGURE 7: Fluid temperature in output from exchangers during 5 years.

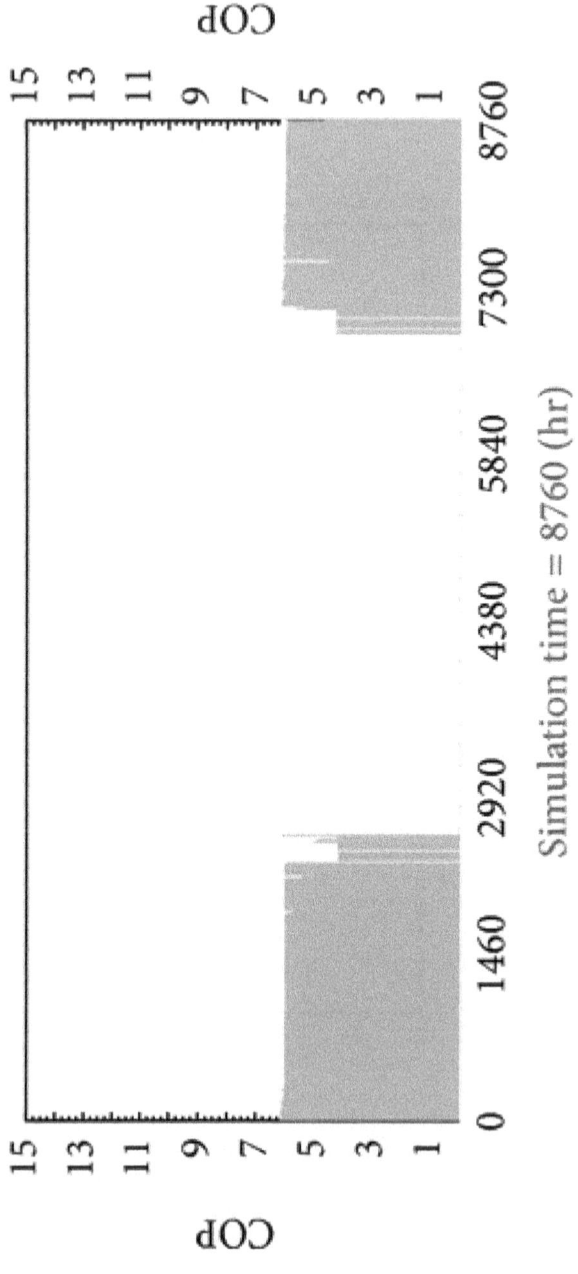

FIGURE 8: COP of heat pump in the first year (during winter time).

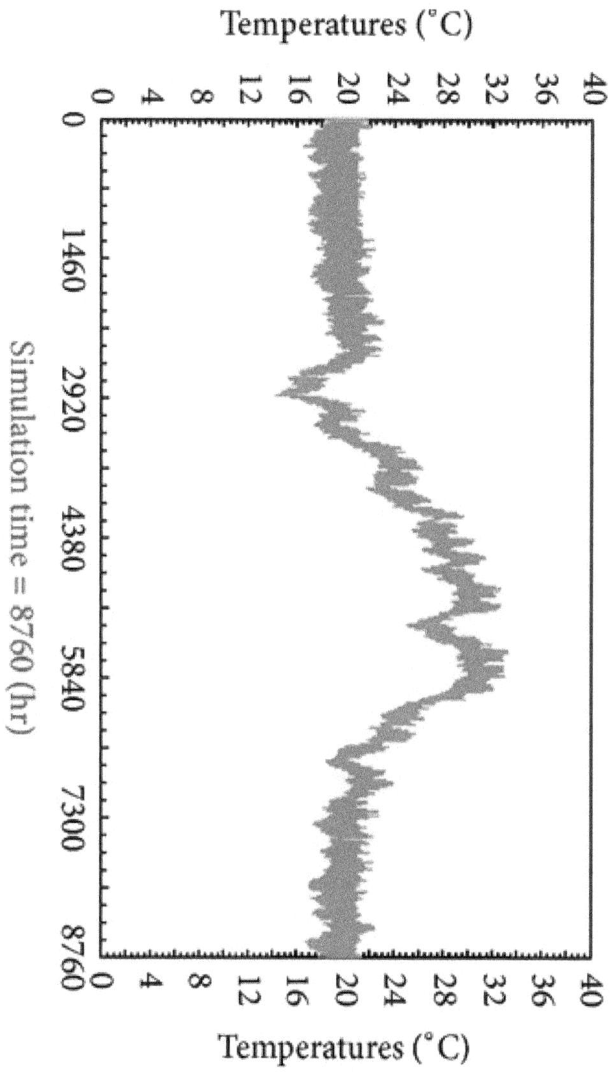

FIGURE 9: Air temperature in the apartment.

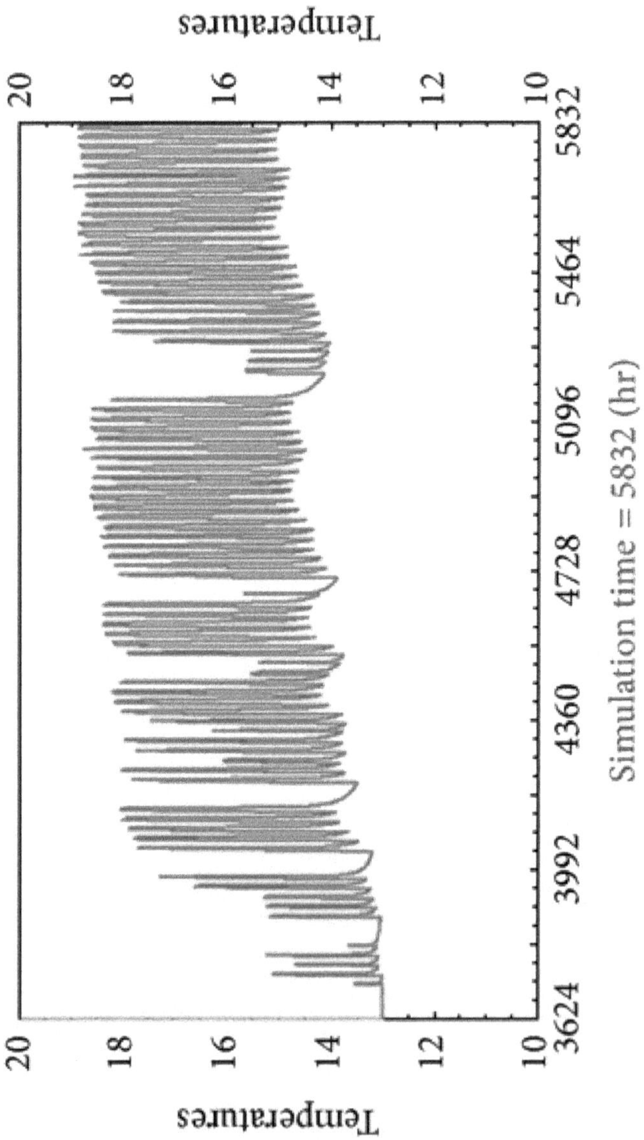

FIGURE 10: Fluid temperature in input in radiant floor in summer (natural cooling).

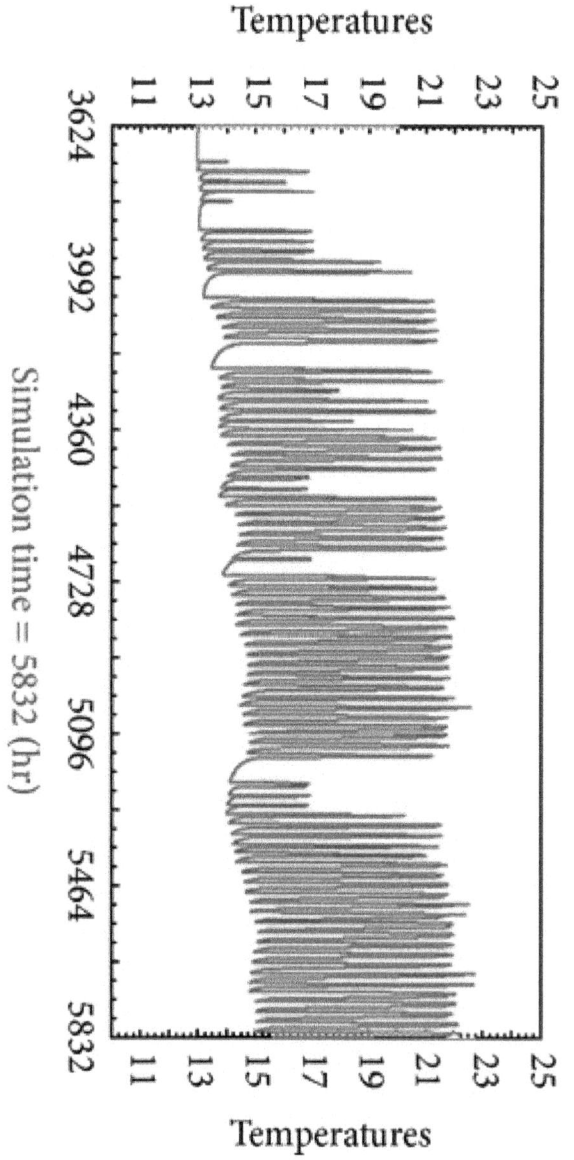

FIGURE 11: Fluid temperature in output from radiant floor in summer (natural cooling).

It can be seen in Figure 6 that during the first year the minimum temperature of the fluid in output from the exchangers is 6°C. In Figure 7 we can see that the same temperature at five years is 5°C. The temperature of fluid in input in the heat pump is in the allowable range (−5, +25°C).

The COP of the heat pump is represented in Figure 8. Note that the performance is good because COP is always more than 4. Moreover, it resulted that the minimum temperature of the fluid in output from the heat pump is 1,5°C. The heating due to the ground is 4,5°C. Air temperature in the apartments (represented in Figure 9) ensure comfort conditions.

The fluid temperature in input in the radiant floors resulted 19°C at the end of the summer, while fluid temperature in output at the end of the summer was 22°C as shown in Figures 10 and 11.

9.4 CONCLUSION

In order to refine the sizing of the system, the TRNSYS 17 software is used. The simulation of the system over a period of 5 years shows the best configuration for the exchangers: 14 vertical heat exchangers in series with distance of 10 m and length of 100 m.

The results show that the system works properly, because

1. the temperature in the apartments ensures comfort conditions;
2. the ground is not subject to overexploiting;
3. high heat pump efficiency.

Simulation with software TRNSYS allows to refine the sizing of the system and accordingly to reduce the initial costs for the ground connection—in particular for drilling and probes that are about the 35% of the total costs—and the operating costs.

REFERENCES

4. S. J. Self, B. V. Reddy, and M. A. Rosen, "Geothermal heat pump systems: status review and comparison with other heating options," Applied Energy, vol. 101, pp. 341–348, 2013.

5. E. Barbier, "Geothermal energy technology and current status: an overview," Renewable and Sustainable Energy Reviews, vol. 6, no. 1-2, pp. 3–65, 2002.
6. M. Carlini and S. Castellucci, "Efficient energy supply from ground coupled heat transfer source," in Proceedings of the International Conference on Computational Science and Applications (ICCSA '10), vol. 6017 of Lecture Notes in Computer Science, part 2, pp. 177–190, Springer, Fukuoka, Japan, 2010.
7. J. W. Lund, D. H. Freeston, and T. L. Boyd, "Direct application of geothermal energy: 2005 worldwide review," Geothermics, vol. 34, no. 6, pp. 691–727, 2005.
8. A. M. Omer, "Ground-source heat pumps systems and applications," Renewable and Sustainable Energy Reviews, vol. 12, no. 2, pp. 344–371, 2008.
9. P. Bayer, D. Saner, S. Balay, L. Rybach, and P. Blum, "Greenhouse gas emission savings of ground source heat pump systems in Europe: a review," Renewable and Sustainable Energy Reviews, vol. 16, no. 2, pp. 1256–1267, 2012.
10. M. Carlini and S. Castellucci, "Modelling tha vertical heat exchanger in thermal basin," in Proceedings of the International Conference on Computational Science and Applications (ICCSA '11), vol. 6785 of Lecture Notes in Computer Science, part 4, pp. 277–286, Springer, Santander, Spain, 2011.
11. M. Carlini and S. Castellucci, "Modelling and simulation for energy production parametric dependence in greenhouses," Mathematical Problems in Engineering, vol. 2010, Article ID 590943, 28 pages, 2010.
12. Y. Bi, X. Wang, Y. Liu, H. Zhang, and L. Chen, "Comprehensive exergy analysis of a ground-source heat pump system for both building heating and cooling modes," Applied Energy, vol. 86, no. 12, pp. 2560–2565, 2009.
13. J. Mazo, M. Delgado, J. M. Marin, and B. Zalba, "Modeling a radiant floor system with Phase Change Material (PCM) integrated into a building simlation tool: analysis of a case study of a floor heating system coupled to a heat pump," Energy and Buildings, vol. 47, pp. 458–466, 2012.
14. C. Cattani, S. Chen, and G. Aldashev, "Information and modeling in complexity," Mathematical Problems in Engineering, vol. 2012, Article ID 868413, 4 pages, 2012.
15. S. Y. Chen, H. Tong, and C. Cattani, "Markov models for image labeling," Mathematical Problems in Engineering, vol. 2012, Article ID 814356, 18 pages, 2012.
16. VDI Technical Division Energy Conversion and Application, Thermal Use of the Underground—Ground Source Heat Pump Systems, 2001.
17. M. Carlini, T. Honorati, and S. Castellucci, "Photovoltaic greenhouses:comparison of optical and thermal behaviour for energy savings," Mathematical Problems in Engineering, vol. 2012, Article ID 743764, 10 pages, 2012.
18. M. Carlini, S. Castellucci, M. Guerrieri, and T. Honorati, "Stability and control for energy production parametric dependence," Mathematical Problems in Engineering, vol. 2010, Article ID 842380, 21 pages, 2010.
19. Solar Energy Laboratory, University of Wisconsin, TRNSYS 17 Documentation.

CHAPTER 10

Recovery of Sewer Waste Heat vs. Heat Pumps Using Borehole Geothermal Energy Storage for a Small Community Water Heating System: Comparison and Feasibility Analysis

SHAHRYAR GARMSIRI, SEAMA KOOHI, AND MARC A. ROSEN

10.1 INTRODUCTION

Space heating, cooling and domestic hot water supply represents the biggest share of energy in residential buildings [1]. There are a number of uses of hot water in buildings, including showers, sinks, dishwashers, clothes washers and others; the waste water retains a significant portion of its initial energy that could be recovered. By removing the waste heat it is possible to reduce the use of fossil fuels for heating water and subsequently reducing greenhouse gas emissions to the atmosphere as a result. The systems described in this section will be used for the analysis of heat recovery system. Figure 1 illustrates the process of the heat recovered and in combination with a heat pump to increase the temperature of the recovered heat for space heating and domestic hot water supply.

It is also possible to extract heat using a closed-loop ground couple heat pump system that relies on heat exchanges to reject or extract heat from the ground. This heat exchanger consists of a borehole in which a U-tube pipe is inserted. The borehole is usually filled with a grout to enhance heat transfer and protect underground aquifers as illustrated in Figure 2.

Alternatively, solar energy is an option for space heating and domestic water heater applications. Many studies have been performed in regards to the solar thermal energy collection as a stand-alone system or the combination of solar thermal energy and ground source heat pump system for domestic hot water heating purposes. In this paper the focus will be on economic and environmental benefits of heat recovery from sewage water in comparison to ground source heat pump systems.

In this paper the ground source heat pump system used is the underground Borehole Thermal Energy Storage (BTES) system that is installed at the University of Ontario Institute of Technology (UOIT) in Oshawa, Ontario, Canada. The BTES's are designed for heating and cooling applications, sometimes with renewable energy in order to minimize greenhouse gas emissions. For this paper the heating portion of this system is considered for the domestic water heater application.

The UOIT BTES system layout as shown in Figure 3 is unique in Canada in terms of the number of holes, capacity, surface area, technology, etc. Large-scale storage systems, comparable to the UOIT system, have been implemented at Stockton College in New Jersey, USA and in Sweden [2]. The conditions for heating the load water leaving the borehole that is 198 m deep to reach the approximate 52°C (126°F) temperature has a heat pump COP of about 3.5 [3]. The system is capable of producing nearly 1,386 MW of energy for heating [3].

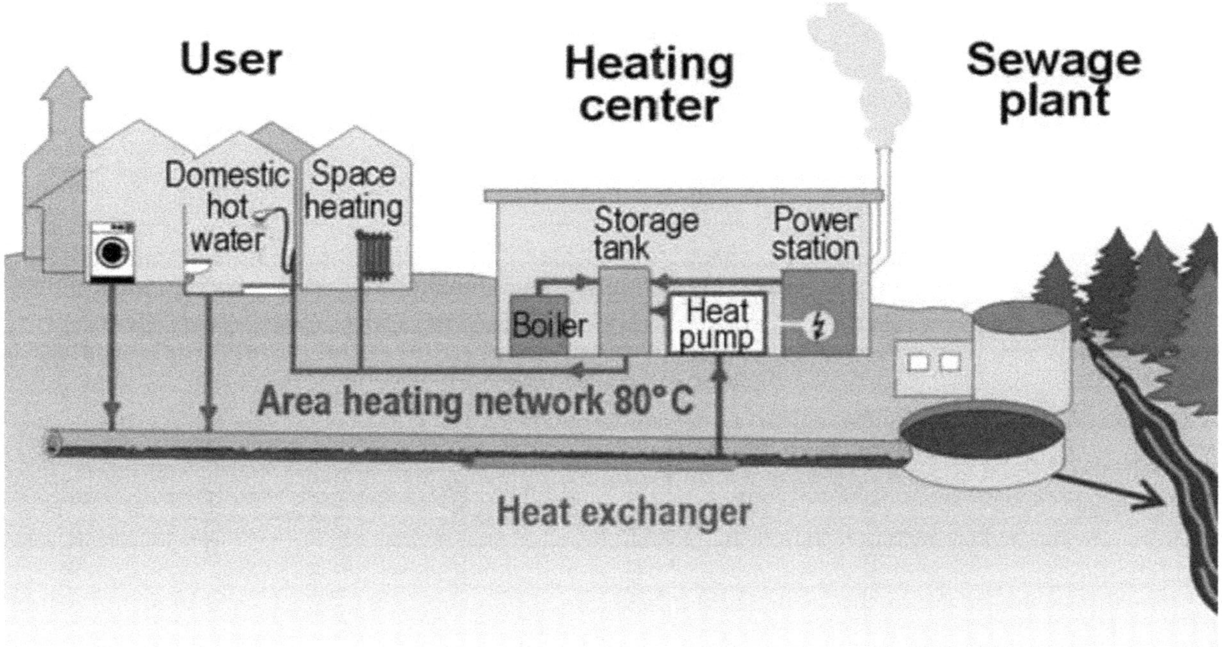

FIGURE 1: Heat recovery from sewage waste heat.

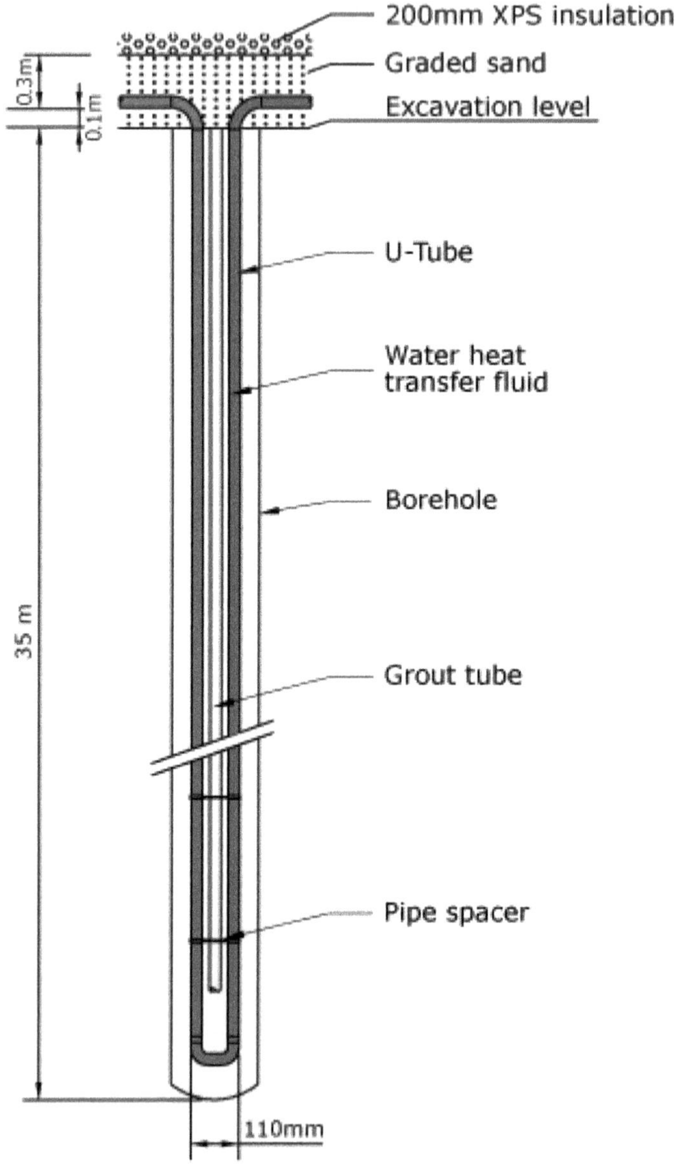

FIGURE 2: UOIT U-tube pipe layout in a borehole.

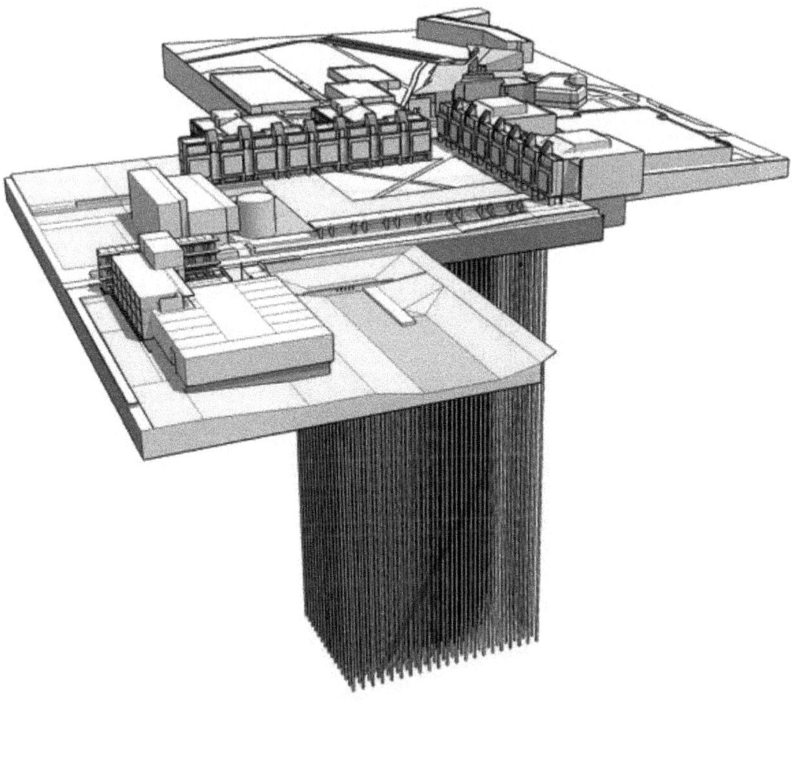

FIGURE 3: UOIT BTES system layout.

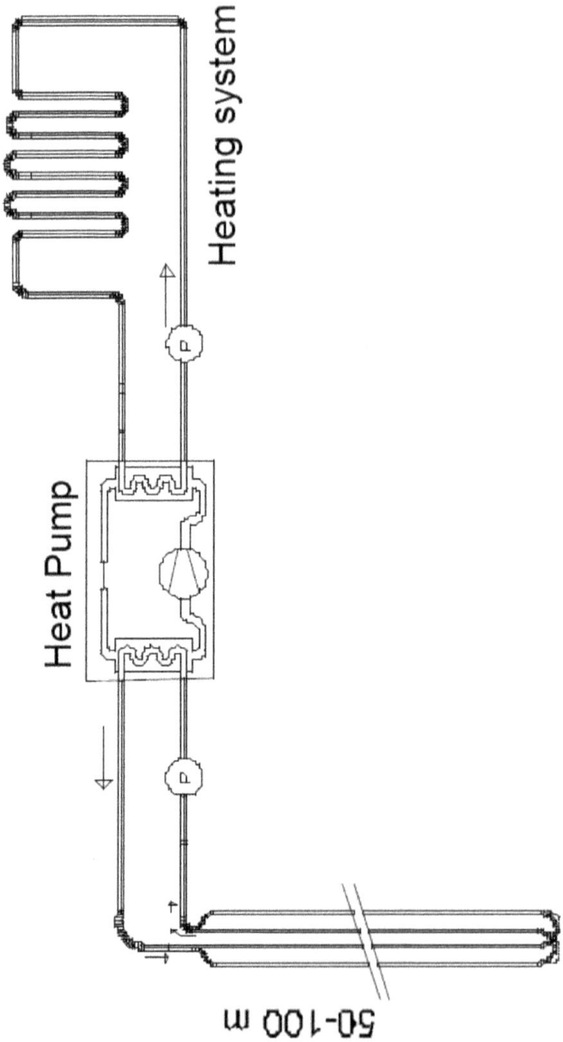

FIGURE 4: Schematic of a ground source heat pump.

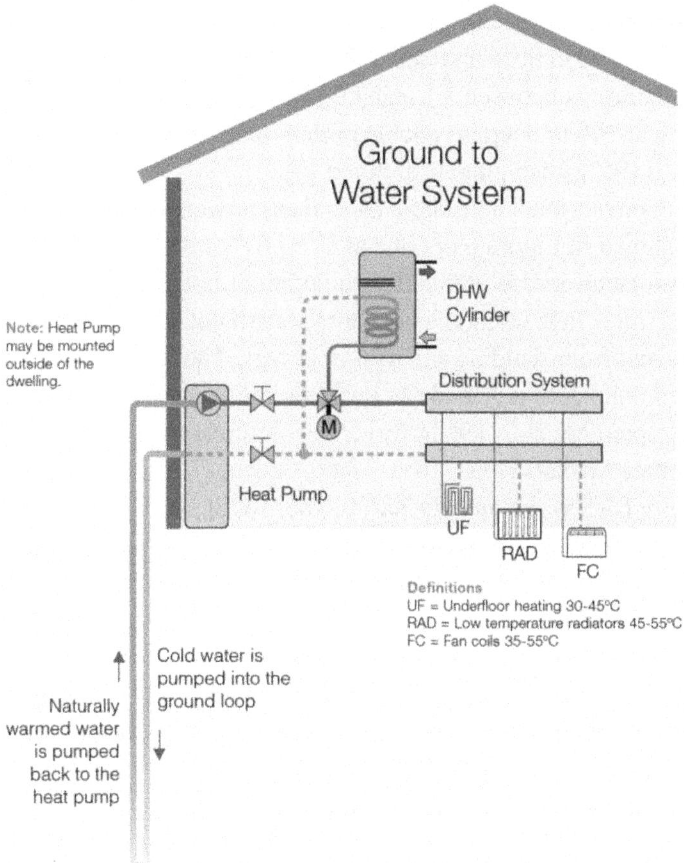

FIGURE 5: Shallow ground source borehole and heat pump system layout.

The UOIT's BTES system is large and costly; however there is a smaller and less expensive type of ground heat source heat exchanger that is known as Shallow Geothermal Energy. The basic principle of a ground source heat pump is shown in Figure 4 and system layout in Figure 5 [4]. Heat can be extracted from the ground at a relatively at low temperatures, transferring the heat through the heat pump to increase the temperature to be useful for space heating [4]. The ground source heat pumps are typically 50 – 100 m deep and intended for closed-loop applications have heating COP ratings between 3.1 and 4.9 [5]. The typical Shallow Geothermal Energy at 100 m deep is capable of delivering a load water of approximately 35°C (95°F) using the heat pump [6].

Conventional household natural gas powered storage tank water heaters as shown in Figure 6 operate at 67%–69% efficiency [7] and water heater manufacturers recommend the maximum hot water temperature supplied to fixtures in residential occupancies shall not exceed 49°C (120°F) to help prevent from scalding and to save energy. The water heater manufacturers' recommendation is made as stated by the Ontario Building Code amended in September 2004 [8], however the requirement exempts dishwashers and clothes washers.

A BTES system similar to the UOIT BTES and shallow ground source BTES are compared simultaneously with the sewer heat exchange method. This is done to determine if using any of the domestic water heating systems, either stand-alone or in combination with other systems, has economic or environmental benefits for municipal water heating purposes. This study is performed in the pursuance of lowering the consumption of natural gas for municipal water heating purposes to reduce CO_2 emissions.

10.1.1 SYSTEM OVERVIEW

There are several options to recover the heat embedded in waste water. These systems portrayed in Figures 1 to 5 shows the possibilities of creating systems for municipal water pre-heating. The heat content of waste water from households can be recovered within houses, this is considered as a small scale application, and may be costly. The heat content may also be recovered from the sewer, considered as medium scale application, or at waste water treatment plants for large scale applications [9]. In Figure 7, the schematic of these systems are outlined. Section A in Figure 7 shows the schematic of a sewage water heat recovery system. The heating water enters the sewer in a counter flow direction of the sewer water flow, absorbs heat and transfers the heat to municipal water line through a heat exchanger. This preheated municipal water then enters the domestic water heater and is heated to the desired temperature for the user. Section B of Figure 7, shows a typical BTES system. In this system, the heat exchanger water passes through the ground loop, absorbs heat and transfers the heat to the municipal water in the heat exchanger. Section C of Figure 7 shows the combination of sewer heat and BTES system for domestic water heating. In this system, the municipal water is preheated in two stages by flowing through a heat exchanger in connection with the sewer system and a second heat exchanger connected to the BTES. The preheated municipal water can then flow to the domestic water heater for additional heat top-off.

In this report, a simple counter flow heat exchanger with stainless steel tubing is considered. A counter flow heat exchanger is considered the ideal type among heat exchangers for the recovery of waste heat energy [10]. The performance of heat exchangers usually deteriorates over time as a result of accumulation of deposits on the tube surfaces. The layers of deposits represent additional resistance to heat transfer and decrease heat transfer rate. The most common fouling is the precipitation of solid deposits in a fluid on heat transfer surfaces. Stainless steel is used due to the limiting factors of fouling formation from the sewer sludge; it is less resistant compared to aluminium tubing [11]. The length of the tube in the counter flow heat exchanger can be evaluated using Equation 1.

$$A_s = \pi DL \quad (1)$$

where, D is the tube diameter and L is the tube length. For a typical stainless steel tube counter flow heat exchanger, a tube diameter of 5 cm is selected with a surface area of 5 m^2. The surface area is estimated for this case and the length is calculated to be 33 m.

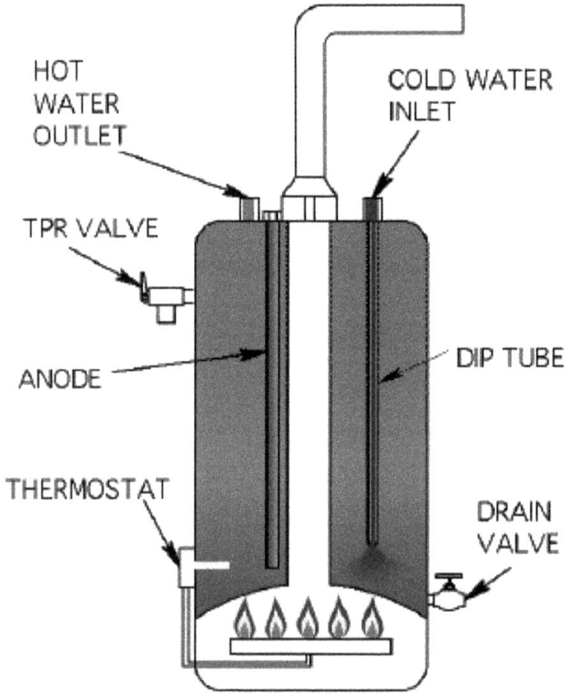

FIGURE 6: Typical household natural gas powered domestic water heater

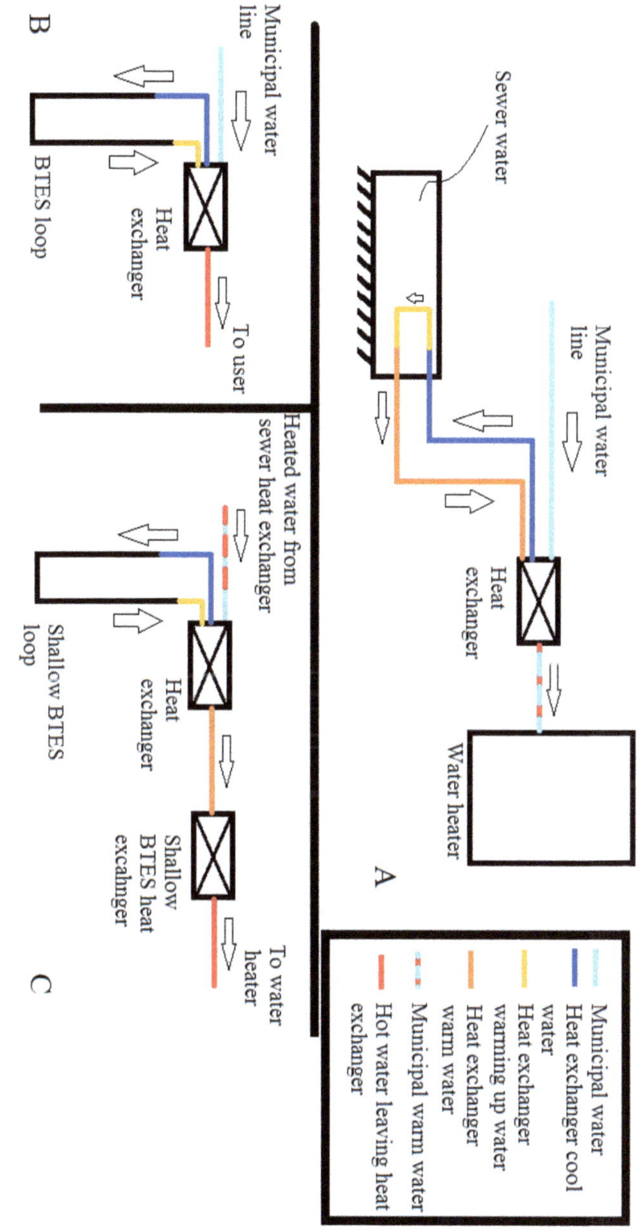

FIGURE 7: Schematic of heat recovery systems.

10.2 FORMULATION

Usually, the modelling of waste water heat recovery presents several obstacles such as acquiring and generating appropriate input data based on highly variable water usage statistics. Since waste water heat has not been considered as a potential source of heat for domestic water heating applications, there are very few statistics available for waste water temperature data. In this study, data related to flow rate and temperature of the waste water in the sewer system of city of Oshawa, Ontario, has been considered. Similarly, the Borehole Thermal Energy Storage (BTES) from University of Ontario Institute of Technology has been considered for this analysis due to availability of information and the university size making it an ideal community size for such application. In some studies, the average sewer temperature considered for calculation is 18°C (64°F) [12]. Oshawa's Storm Sewer Use By-law 95-95, updated in 2013, indicates a 40°C (104°F) as the maximum temperature limit for storm sewer discharge [13].

The temperatures for waste water treatment range from 25°C (77°F) to 35°C (95°F) [14]. In general, biological treatment activity accelerates in warm temperatures and slows in cool temperatures but extreme hot or cold can stop treatment processes altogether. Thus, due to the variant temperature of the sewage water, a range of 18°C (64°F) to 26°C (79°F) is used for the calculations.

Multiple sources indicate the municipal water temperature is less than or equal to 15°C (59°F) [15, 16]. This temperature is adopted from the guidelines of Canadian drinking water [17]. The rational is that at temperatures above 15°C (59°F), the growth of nuisance organisms in the distribution system can lead to the development of unpleasant tastes and odors. Thus, the cold water temperature of 15°C is used for the analysis.

In this study, the heat transfer coefficient is 675 W/m^{2}K [18] for a common tubular counter flow heat exchanger with a double pipe shell and stainless steel tube construction. For this analysis, a small community such as UOIT has been considered that uses approximately 13,627 L (3,600 gallons) of water per day.

The energy input equation is shown below:

$$Q_{in} = (\rho_W V_W c_p \Delta T_W) / \eta_{WH} \quad (2)$$

where, Q_{in} is the daily energy input for heating municipal water, ρ_W is the water density at 15°C (59°F), V_W is the water tank volume, c_p is the specific heat capacity of the municipal water, ΔT_W is the temperature difference between the water entering the municipal water heater and user's desired temperature, and η_{WH} is the domestic water heater efficiency. When the municipal water is preheated before entering the tank, Equation 3 is used to calculate the additional heat required to heat the municipal water to the user's desired temperature (49°C in this study).

TABLE 1: Fixed variables used in the study.

Fixed variables	Values	Units
Tank size	13,627	L
Mass flow rate of cold municipal water (15°C) (59°F)	9.46	kg/s
Mass flow rate of warm water (25°C) (77°F)	163.91	kg/s
C_p	4.18	kJ/Kg°C
Water heater efficiency	68	%
Water density (ρwater)	0.99975	g/m³
Energy to mass conversion (kJ to kJ/m³)	35,435	kJ/m³
Natural gas price	22	cents/m³
Water exiting the water heater ($T_{c,out}$)	49	°C

This conversion rate is outlined in Table 1. The temperature of the municipal water exiting the heat exchanger preheating the water can be determined using the counter flow heat exchanger energy balance as shown in Equation 3. In this analysis, $T_{c,in}$ is the temperature of the cold water from municipal water pipeline and $T_{c,out}$ is the water temperature exiting the heat exchanger. The variable $T_{h,in}$ is the inlet warm water from the sewage, and $T_{h,out}$ is the outlet sewage warm water after it has passed through the heat exchanger [19].

$$Q_{max} = C_{min} (T_{h,in} - T_{c,in}) \quad (3)$$

where C_{min} is the minimum heat capacity value between the heat capacity of the cold water (C_c) and the heat capacity of the hot water (C_h),

$$C_c = m_c C_{p,c} \quad (4)$$

$$C_h = m_h C_{p,h} \quad (5)$$

The outlet temperatures after the heat transfer has taken place are determined by

$$Q_c = C_c (T_{c,out} - T_{c,in}) \quad (6)$$

$$Q_h = C_h (T_{h,in} - T_{h,out}) \quad (7)$$

where Q_c is the heat transferred to the cold water and Q_h is the heat transferred from the hot water.

Equations 6 and 7 can be rearranged to determine the $T_{c,out}$ as illustrated in Equation 8 and to understand how much heat can be transferred for other applications. If a combination system is considered that uses two heat exchangers, then it is possible to calculate the outlet temperature of the municipal water from the first heat exchanger using Equation 8 to calculate the heat transferred through the second heat exchanger. In addition, Equation 9 can be used to determine the heat content of the outlet heat exchanger water temperature and weather it can be reused to capture more heat.

$$T_{c,out} = T_{c,in} + (Q_{max}/C_c) \quad (8)$$

$$T_{h,out} = T_{h,in} + (Q_{max}/C_h) \quad (9)$$

A typical problem in the analysis of a heat exchanger is the performance calculation. Given the inlet conditions to evaluate how the exchanger performs, and determining the outlet temperatures. Using Equation 1, the solution may be reached only by trial-and-error. An alternate approach is the notion of heat exchanger effectiveness, E, as outlined in Equation 10.

For this study it is assumed a 100% heat transfer, making the effectiveness factor a value of one. This is done so to simplify the calculations and by varying the effectiveness value would require further analysis. The purpose of this study is to show the other available options of heating water to reduce the use of natural gas to heat water.

$$E = (\text{actual heat transfer/maximum possible heat transfer}) = (T_{h,in} - T_{h,out}) / (T_{h,in} - T_{c,in}) \quad (10)$$

10.3 RESULTS AND DISCUSSION

In this paper, a set of fixed variables, identified in Table 1, are used to determine the outlet municipal water temperature as it passes through the conventional tank water heater using natural gas. The results in Table 2 show that approximately 2.85 GJ of energy, equivalent to about 80 m^3 of natural gas, is required to heat the 13,627 L (3,600 gallons) of water per day. This results in an annual natural gas cost of $6,567.

The results in Table 2 show the daily natural gas usage for the tank water heater. As it is expected, for cases where the water is preheated before entering the tank water heater, less natural gas is used to heat the water to the user's required temperature (49°C). The annual natural gas cost is calculated based on the natural gas cost per cubic meter in southern Ontario. The following column shows the annual natural gas cost savings resulting from pre-heating the municipal water.

TABLE 2: Results from the analysis.

Description	Temperature range (°C)	Heat exchanger efficiency (%)	Water entering the water heater $T_{c,in}$ (°C)	ΔT (°C)	Daily energy usage Qin eff (GJ)	Natural gas mass equivalent (m^3)	Annual natural gas cost ($/year)	Annual natural gas cost savings ($/year)
Natural gas powered domestic hot water tank	15 - 49	0	15	34	2.85	80	$6,567	$0
Heat recovery from sewer waste water	15 - 18	80	17	32	2.56	75	$6,219	$348
Shallow borehole geothermal system	15 - 35	100	35	14	1.17	33	$2,704	$3,863
UOIT borehole geothermal system	15 - 49	100	NA	0	-	0	$0	$6,567
Combination of sewer waste water and shallow borehole	15 - 49	100	35	14	1.17	33	$2,704	$3,863

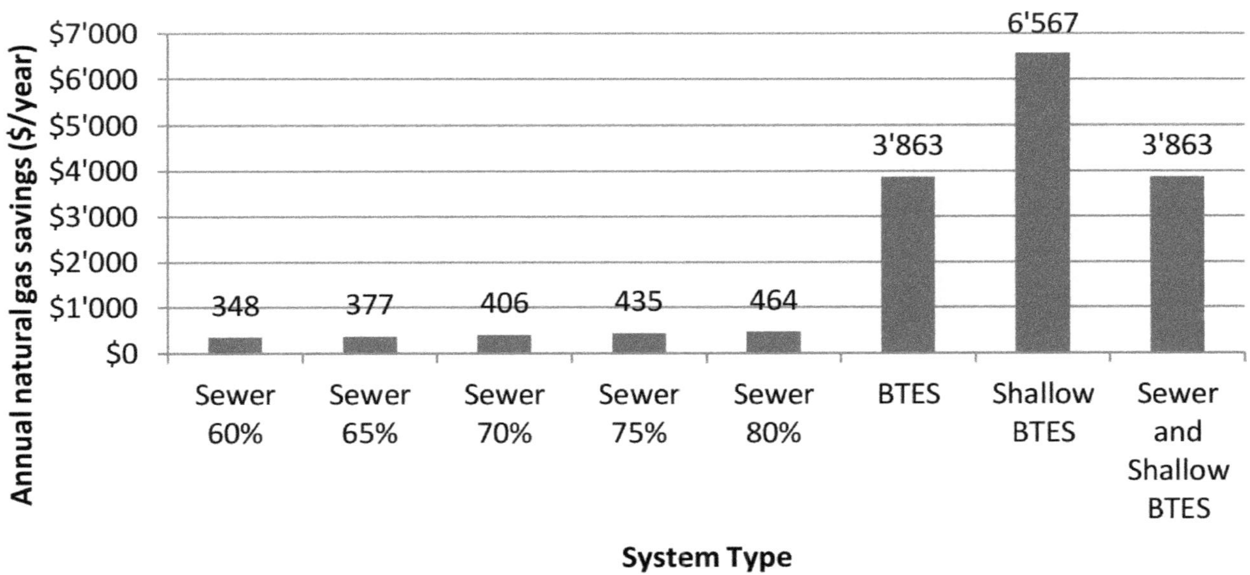

FIGURE 8: Annual natural gas cost savings.

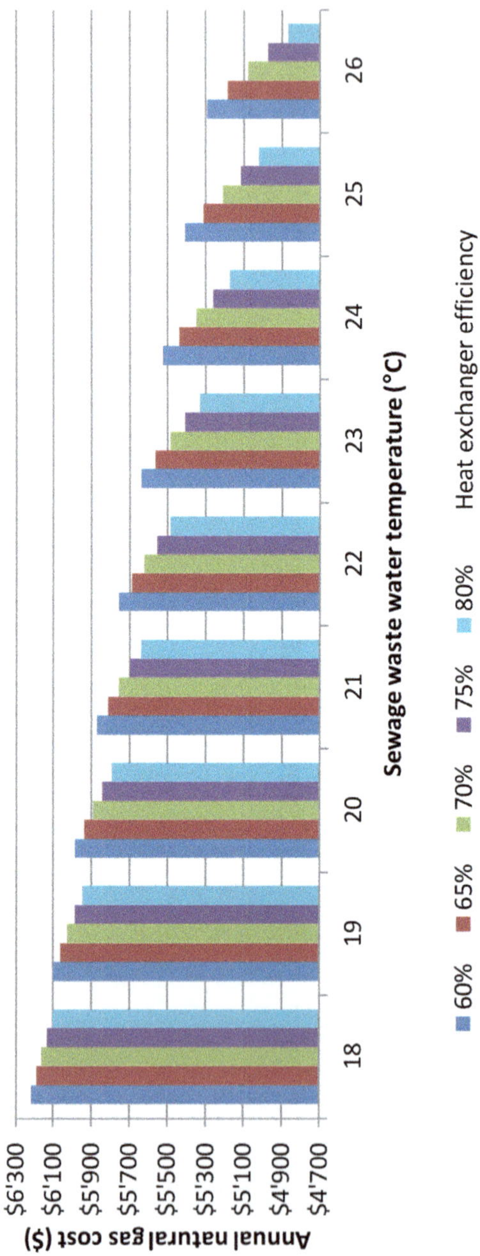

FIGURE 9: Variation of annual natural gas cost with swage waste water temperature and sewer heat exchanger efficiency ranging from 60 - 80%.

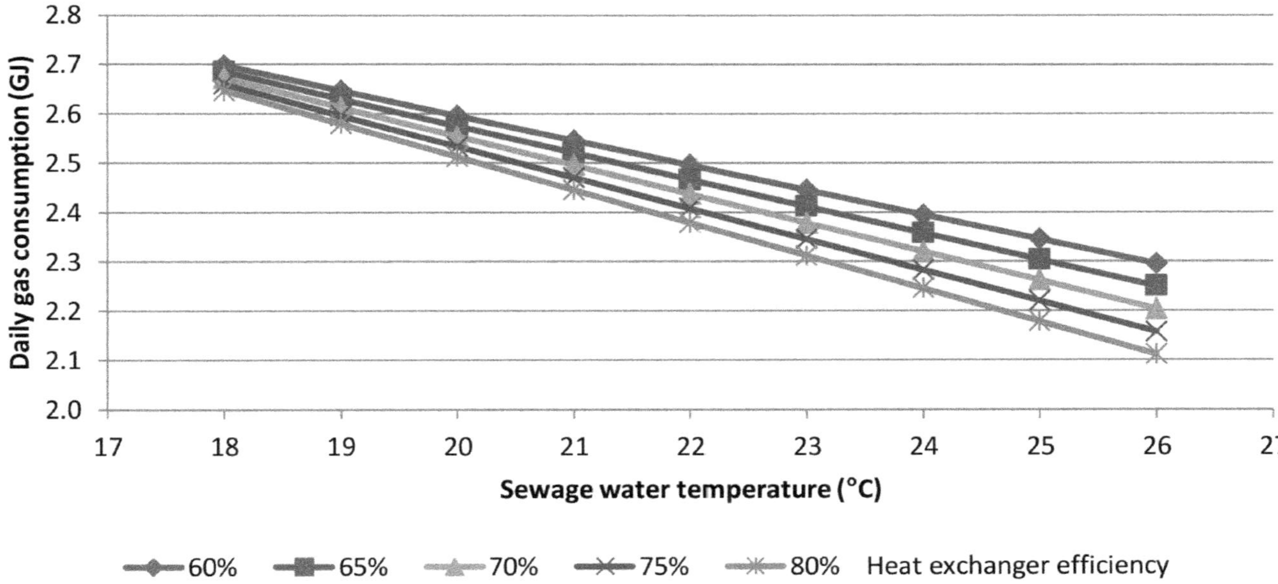

FIGURE 10: Sewer heat exchanger efficiency ranging from 60 - 80% and daily natural gas energy usage.

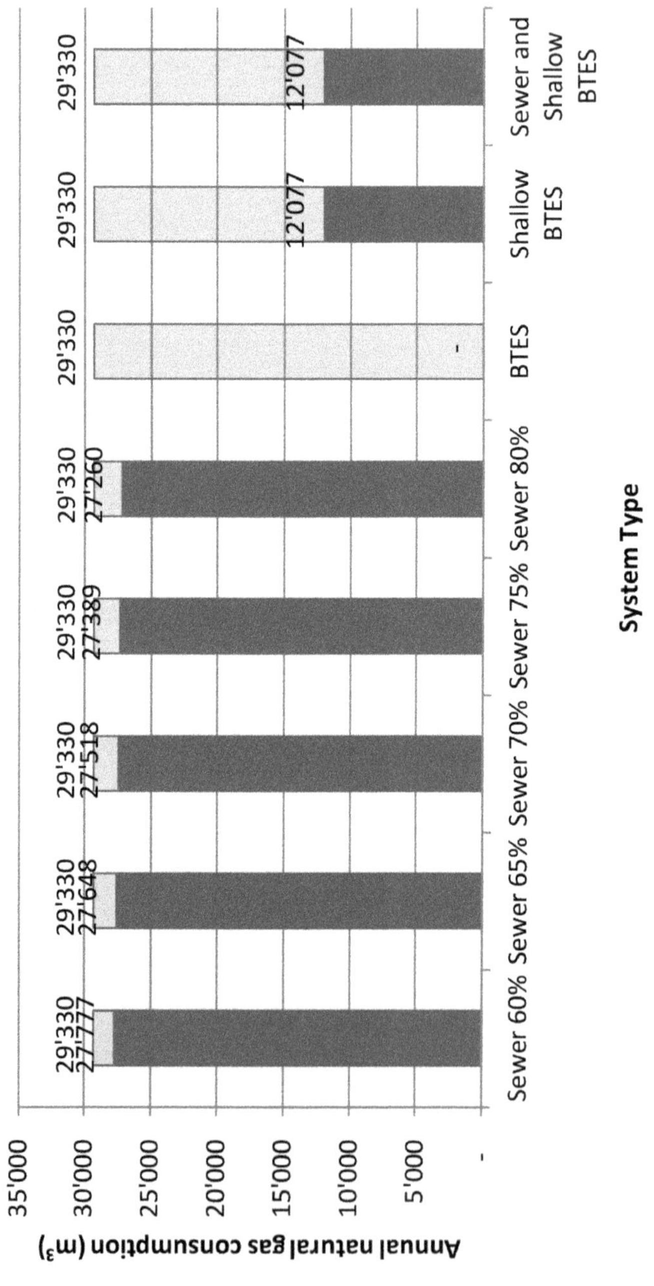

FIGURE 11: Annual natural gas usage per system compared to traditional natural gas domestic hot water method.

It is shown in Table 2 that the water coming out of the heat exchanger is calculated to be 17°C. The amount of saved energy using this method is approximately 0.29 GJ with annual savings of $348 in natural gas. The natural gas cost savings are outlined in Figure 8 for each system. It is notable that the shallow BTES system results in the highest annual natural gas savings. It is also noteworthy to observe the incremental savings in the different heat exchanger efficiencies for the sewer system.

The cost of this simple sewage heat exchanger system is typically around $8,000 [20] and it is possible to calculate the rate of return on investment using a simple payback period method through the calculation of an Internal Rate of Return (IRR). IRR is a rate of return used in capital budgeting to measure and compare the profitability of investments. Accounting for the natural gas savings, the payback period is calculated to be approximately 17 years. However, this payback period decreases as the efficiency of the heat exchanger increases. The optimal point, when comparing heat exchanger efficiency and sewage water temperature in this study, is 80% heat exchanger efficiency and 26°C sewage water temperature. In this case, the simple payback period reduces from approximately 17 to 5 years.

Figure 9 displays the annual natural gas cost for heat exchanger efficiency that is between 60%-80% and various sewage water temperatures based on the maximum allowable discharge temperatures set by the municipalities. Figure 9 show the variation of the annual natural gas cost with heat exchanger efficiency. The higher heat exchanger efficiencies result in higher energy savings and lower natural gas consumption and CO_2 emissions. With higher heat exchanger efficiency, the water leaving the heat exchanger is at a higher temperature requiring less natural gas for the additional heat top-off. Similarly, as the sewer water temperature increases, more savings are observed. Figure 10 shows the daily natural gas energy required for heat top-off after the heat recovery for the sewage system. The heat top off is necessary as the municipal water temperature should reach 49°C to meet the user needs and the sewer system alone is not capable of heating the water to this temperature.

Table 2 also shows the results for the UOIT BTES system. As it is observed, no top-off heat is required, as the system is capable of producing enough heat to bring the municipal water temperature to the desired

49°C. Thus, an annual savings of \$6,567 and prevention of 2.85 GJ/day energy of natural gas is achievable. Although this may appear to be an ideal system, the cost of this system is relatively higher as compared to the other systems and is approximately \$7,000,000 [21] and would take approximately over 100 years to pay back initial cost based on only natural gas usage costs.

In Table 2, it is shown that the shallow ground source BTES system can increase the municipal water temperature to 35°C by recovering heat from the underground source using heat pumps. The annual Natural Gas cost is \$2,704 and, for the 13,627 L (300 gallon) of water use per day application, the typical cost is approximately £11,000 to £15,000 (\$20,000–\$27,500 CAD) [22] with a rate of return of approximately 6 years on this investment.

Lastly, in Table 2, natural gas cost/savings and temperatures of municipal water entering the heat exchanger are displayed for the combination of sewer heat exchanger with a shallow ground source BTES. It is shown that the annual natural gas cost for this system is similar to the shallow ground source BTES system at \$2,704. The municipal water from the sewage heat exchanger enters the shallow BTES heat exchanger at 17°C bringing the municipal water temperature up to the 35°C before entering the water heater for additional heat top-off. It is notable that in this calculation, only the initial investment cost has been considered. This setup is a combination of both sewer heat exchanger and the shallow ground source BTES with a total investment cost of approximately \$35,000. The payback period can be calculated to be approximately 8 years.

Figure 11 summarizes the annual natural gas usage as outlined by the blue bars, for each system and for the combination of the systems in comparison with the traditional domestic hot water system with no preheating system. The pink bars are the total natural gas usage for the traditional domestic hot water system. The difference is the savings in natural gas by using the municipal water preheating systems.

10.4 CONCLUSIONS

The benefits of utilizing waste heat and ground source heat as a means to improve the economic performance and to reduce CO_2 emissions has been analysed. Each system showed potential significant energy savings economically. Furthermore, energy savings would help prevent greenhouse gas emissions into the atmosphere. The energy recovery systems considered in this paper were the sewage warm water heat recovery, borehole thermal energy storage system and a shallow borehole thermal energy storage system. Each system was analysed and compared to the traditional natural gas powered water heating systems currently used in homes. The costs of each system were obtained from various sources and, using only the natural gas savings from each system, a simple payback period had been calculated. The sewage system appeared more economical with higher heat exchanger efficiency and higher sewage water temperature. Although the combination system showed a faster rate of return on investment using the simple payback period method, it was also more costly as compared to the sewer heat exchanger system as an initial investment.

The borehole thermal energy storage system showed no natural gas usage, making it the most environmental beneficial when compared to the other systems in this analysis. When considering the economic factors only, the sewer system is considered as a sustainable and readily available system. This is due to readily available sewer in most areas where the heat exchanger unit can be installed with minimal labour and construction costs.

REFERENCES

1. H. Torio and D. Schmidt, "Development of system concepts for improving the performance of a waste heat district heating network with exergy analysis.," Energy and Buildings, vol. 42, no. 10, pp. 1601-1609, 2010.
2. International Energy Agency, "Energy Conservation through Energy Storage.," 2005. [Online]. Available: http://www.iea-eces.org/success/success.html. [Accessed February 2014].
3. H. Paksoy, "Thermal Energy Storage for Sustainable Energy Consumption, Fundamentals, Case Studies and Design.," Springer, vol. 234, no. 12, pp. 221-228, 2007.
4. B. Sanner, "Shallow geothermal energy.," 1998. [Online]. Available: https://pangea.stanford.edu/ERE/pdf/IGAstandard/ISS/2003Germany/II/4_1.san.pdf. [Accessed March 2014].
5. Natural Resources Canada, "Ground-Source Heat Pumps (Earth-Energy Systems)," 2013. [Online]. Available: https://www.nrcan.gc.ca/energy/publications/efficiency/residential/heating-heatpump/6833. [Accessed March 2014].

6. J. Hanova, H. Dowlatabadi and L. Mueller, "Ground source heat pump systems in Canada.," 2007. [Online]. Available: http://www.greenerg.com/documents/Canada_GSHP.pdf. [Accessed March 2014].
7. Energy Star, "Energy star water heater program.," 2014. [Online]. Available: http://www.fortisbc.com/NaturalGas/Homes/Offers/EnergyStarTanklessWaterHeaterProgram/Pages/default.aspx. [Accessed February 2014].
8. Canadian Plumbing Code, "Canadian Plumbing Code – Maximum Hot Water Temperature," February 2008. [Online]. Available: http://www.cashacme.com/_images/pdf_downloads/products/. [Accessed February 2014].
9. J. Frijns, J. Hofman and M. Nederlof, "The potential of (waste) water as energy carrier.," Energy Conversion and Management , vol. 65, no. Global Conference on Renewable energy and Energy Efficiency for Desert Regions 2011 "GCREEDER 2011", pp. 357-363, 2013.
10. S. Hsieh and D. Huang, "Thermal performance and pressure drop of counter-flow and parallel-flow heat-pipe heat exchangers with aligned tube rows.," Heat recovery systems and CHP, vol. 8, no. 4, pp. 343-354, 1988.
11. Wolverine Engineering, "Wolverine Engineering data book 2.," 2001. [Online]. Available: http://www.wlv.com/products/databook/databook.pdf. [Accessed March 2014].
12. S. Cipolla and M. Maglionico, "Heat recovery from urban wastewater: Analysis of the variability of flow rate and temperature.," Energy Procedia, vol. 45, no. ATI 2013 - 68th Conference of the Italian Thermal Machines Engineering Association, pp. 288-297, 2014.
13. Oshawa's Commisioner, Development Services Department, "Update of storm sewer use by-law 95-95. Public Record, Oshawa, Ontario.," 2013. [Online]. Available: http://www.oshawa.ca/agendas/development_services/2013/03-25/DS-13-66-UpdateStorm-Sewer-Use-Bylaw.pdf. [Accessed February 2014].
14. Northern Arizona University, "Onsite wastewater demonstration project, characteristics of residential wastewater.," 1997. [Online]. Available: http://www.cefns.nau.edu/Projects/WDP/resources/Characteristics.htm#references. [Accessed March 2014].
15. Ministry of Environment, "Water Quality Guidelines for Temperature.," 2001. [Online]. Available: http://www.env.gov.bc.ca/wat/wq/BCguidelines/temptech/temperature.html. [Accessed March 2014].
16. Health Canada, "Environmental and Workplace Health.," 2009. [Online]. Available: http://www.hc-sc.gc.ca/ewh-semt/pubs/water-eau/temperature/index-eng.php. [Accessed February 2014].
17. Service Ontario, "Safe Drinking Water Act, 2002.," 2002. [Online]. Available: http://www.e-laws.gov.on.ca/html/regs/english/elaws_regs_030170_e.htm. [Accessed March 2014].
18. Engineering Toolbox, "Overall heat transfer coefficients in some common heat exchanger constructions - tubular, plate and spiral.," 2014. [Online]. Available: http://www.engineeringtoolbox.com/heat-transfer-coefficients-exchangers-d_450.html. [Accessed March 2014].
19. Y. A. Cengel, Heat and Mass Transfer: A Practical Approach, New York: McGraw Hill, 2007.

20. Compass Resource Management, "Surrey sewer heat recovery study," 2009. [Online]. Available: http://www.surrey.ca/files/SurreyPreliminarySewerHeatRecoveryDraftReport.pdf. [Accessed March 2014].
21. B. Rezaie, B. Rezaie and M. Rosen, "Economic and CO2 emissions comparison of district energy systems using geothermal and solar energy resources.," in 3rd world sustainability forum, web conference, Toronto, ON, 2013.
22. Energy Cost, "Ground source heat pumps.," 2014. [Online]. Available: http://www.energysavingtrust.org.uk/Generating-energy/Choosing-a-renewabletechnology/Ground-source-heat-pumps. [Accessed March 2014].

PART V

EFFICIENCY AND FEASIBILITY OF RENEWABLE ENERGIES

CHAPTER 11

Development of a Wind Directly Forced Heat Pump and Its Efficiency Analysis

CHING-SONG JWO, ZI-JIE CHIEN, YEN-LIN CHEN,
AND CHAO-CHUN CHIEN

11.1 INTRODUCTION

Heavy energy consumption and environmental pollution have become very serious issues in today's world. Finding and developing new alternative energies has become a very urgent issue. Although an alternative fuel could be found to replace the petroleum oil and resolve the energy crisis problem, the gas waste would pollute the environment, which should be taken into account in the energy issues and environmental protection issues. It is thus a very important subject to develop a new kind of energy [1] that would produce less waste gas. Green energies that can replace fuel energy include wind energy, solar energy, biomass energy, and tidal energy, with the characteristic of being free, natural, free of pollution problems, and inexhaustible. The green energies are best suited to the contemporary needs. Currently, most of the focuses are on the development of wind energy [2, 3] and solar energy [4, 5]. The cost of developing wind force energy is much lower than that for the solar energy and that is the major advantage of wind force energy. For instance, the advanced countries, such as Germany, Denmark, the UK, and the USA have invested in the development of wind force energy [6]. The technology of converting wind force energy to electric energy has been maturely developed with many of the development results being applied to the power utilization in the people's livelihood, commerce, and industry [7, 8].

A place with abundant wind energy is the essential condition for the wind turbine. Taiwan is an island country surrounded by sea with many small islands, which is very suitable for wide applications of the wind power generation [9]. Since it is very cold in the winter and very hot in the summer in the seaside environments, it is even suitable to promote the method of driving the heat pump system to produce cool water and hot water directly with the wind force. Figure 1 shows the statistics of the mean wind speeds of several local areas in Taiwan. As can be seen from Figure 1, the mean average wind speeds during all seasons are over 4 m/s [10]. This study proposes a new technique to directly adopt the wind force to drive heat pump systems, which can effectively reduce the energy conversion losses during the processes of wind force energy converting to electric energy and electric energy converting to kinetic energy.

11.2 OVERVIEW OF THE PROPOSED WIND DIRECTLY FORCED HEAT PUMP

The common wind turbines are in two types, namely, the horizontal type and vertical type, with respective advantages and disadvantages. The vertical type is used in the medium- and small-sized wind-driven generators, while the horizontal type is adopted in the large- and medium-sized wind-driven generators. We propose a new technique to directly adopt the wind force to drive heat pump systems, which can effectively reduce the energy conversion losses during the processes of wind force energy converting to electric energy and electric energy converting to kinetic energy. The power transferring the wind force energy indirectly to the heat pump system by the traditional method is calculated with the formula as

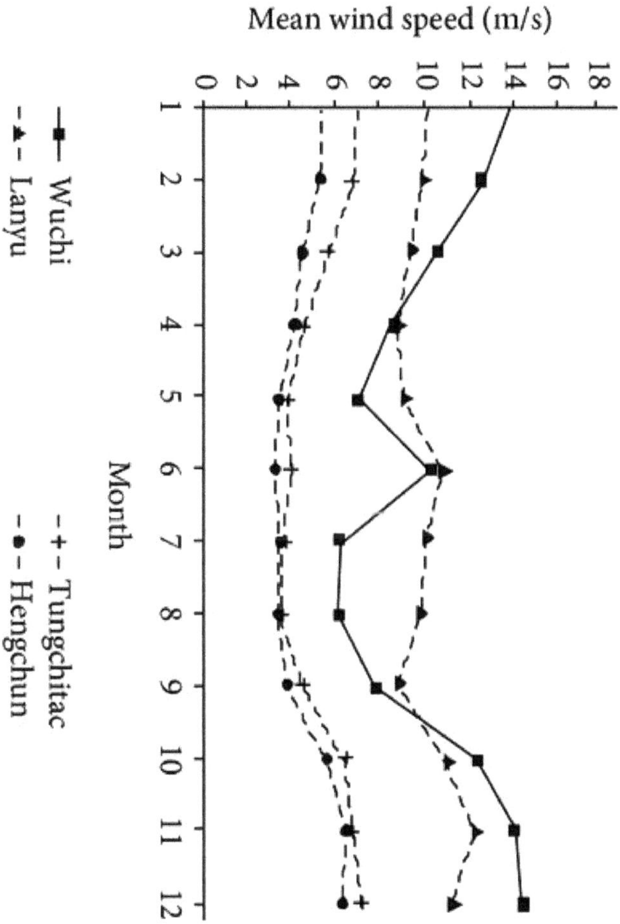

FIGURE 1: Monthly mean wind speed for the four towns in Taiwan.

$$p_{hp} = \eta_{\omega} \times \eta_{T1} \times \eta_{G} \times \eta_{T2} \times \eta_{M} \times P_{\omega} = 42.19\% \times P_{\omega} \quad (1)$$

where P_{ω} is the power of wind force; p_{hp} is the power when reaching the heat pump system; η_{ω} is the efficiency of wind turbine; η_{T1} is the efficiency of the transmission between the wind turbine and generator; η_{G} is the efficiency of generator; η_{T2} is the transfer efficiency between the generator and electromotor; and is the efficiency of electromotor.

According to the literatures, η_{ω} is around 0.59 according to the Betz Law; η_{T1}, the efficiency of the transmission, is around 0.95 usually; η_{G}, the efficiency of generator, is around 0.8 [11]; η_{T2}, the transfer efficiency between the generator and electromotor is around 0.97; and η_{M}, the efficiency of electromotor, is around 0.97. A large quantity of energy losses are generated in the process of converting the wind force energy to electric energy and then the electric energy to mechanical energy driving the compressor to produce the cool energy or hot energy [12]. Using (1), we could get that the total theoretical efficiency of wind energy transfered to thermal energy is 42.19%.

In this study, the wind force could drive the heat pump system directly so that the energy conversion loss could be reduced and the efficiency could be further increased. If an efficient method of the heat pump driving system to generate cool or hot energy directly with the wind force proposed in this study is adopted, then the costs of the generator and generator equipments can be effectively reduced, and the efficiency of converting the wind force energy into the mechanical energy serving a double purpose can also be significantly improved. The power transferring of the wind force energy directly to the heat pump system by the proposed method is calculated with the formula:

$$p_{hp} = \eta_{\omega} \times P_{\omega} \quad (2)$$

This study tests in the four average wind speeds, namely, 3 m/s, 4 m/s, 5 m/s, and 6 m/s, are taken as the basis for performance analysis to verify the feasibility on driving the heat pump system to produce cool water and hot water directly with the wind force.

11.3 ANALYSIS OF WORKING EFFICIENCY

This study analyzes the working efficiency of the cooling and heating achieved by driving the heat pump system directly with the vanes of wind turbine. The theoretical analysis is detailed in the following subsections.

11.3.1 THE WIND ENERGY

The wind turbine collects the kinetic energy of the wind force through the vanes and outputs the energy through the drive shaft, which is expressed as follows:

$$E = 0.5\rho A V^3 t \tag{3}$$

where ρ denotes the air density; A is the area of wind turbine vanes; V represents the wind speed; t is the wind turbine operation time.

11.3.2 THE HEAT PUMP ENERGY

The cooling capacity of the heat pump and the heat output to reduce the temperature of the water in the chiller water tank from t_{c2} to t_{c1} is expressed as follows:

$$Q_c = m \times s \times \Delta t_c = m \times s \times (t_{c2} - t_{c1}) \tag{4}$$

where m is the water storage weight of the chiller water tank; s is the specific heat of water; Δt_c is the water temperature difference of the chiller water tank.

The heating capacity of the heat pump and the heat being absorbed to increase the water temperature in the hot water tank from t_{h1} to t_{h2} could be formulated as follows:

$$Q_h = m \times s \times \Delta t_h = m \times s \times (t_{h2} - t_{h1}) \tag{5}$$

where m is the water storage weight of the hot water tank; s is the specific heat of water; Δt_h is the water temperature difference of the hot water tank.

11.3.3 THE EFFICIENCY

The cooling efficiency η_c or heating efficiency η_h of the heat pump system driven directly by the vanes of wind turbine could be formulated as follows:

$$\eta_c = (Q_c/E) \times 100\%$$

$$\eta_h = (Q_h/E) \times 100\% \tag{6}$$

where Q_c is the cooling capacity of heat pump; Q_h is the heating capacity of heat pump; E is the energy output by the drive shaft of wind turbine.

11.4 THE PROPOSED WIND DIRECTLY FORCED HEAT PUMP

The experimental equipment of the proposed system consisted of three major parts, namely, the wind turbine, transmission, and heat pump system. The detailed specification, structure, and experimental process are shown as follows.

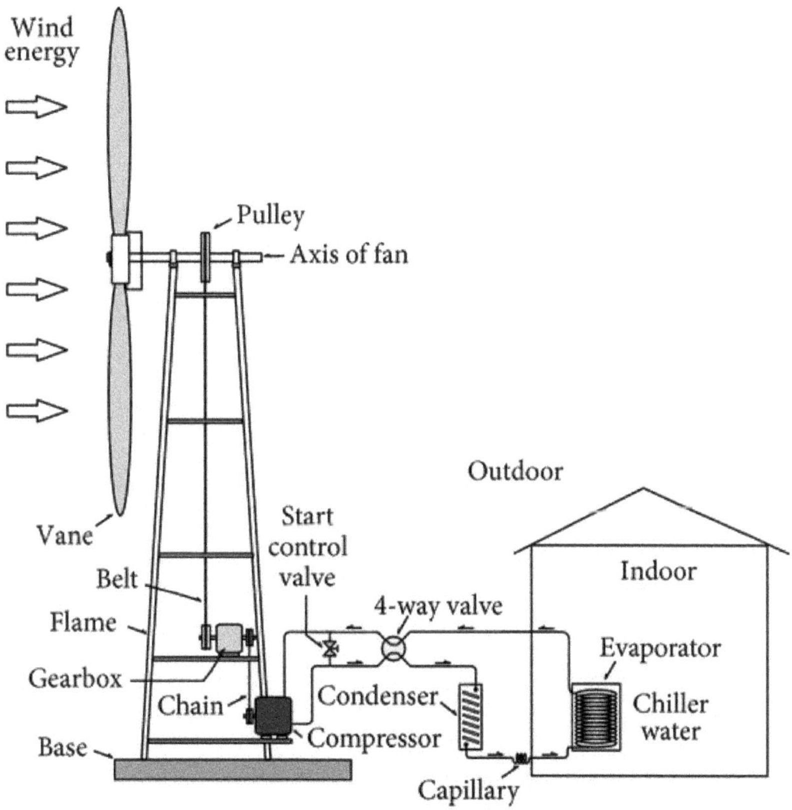

FIGURE 2: Summer model (chiller) of wind directly forced heat pump.

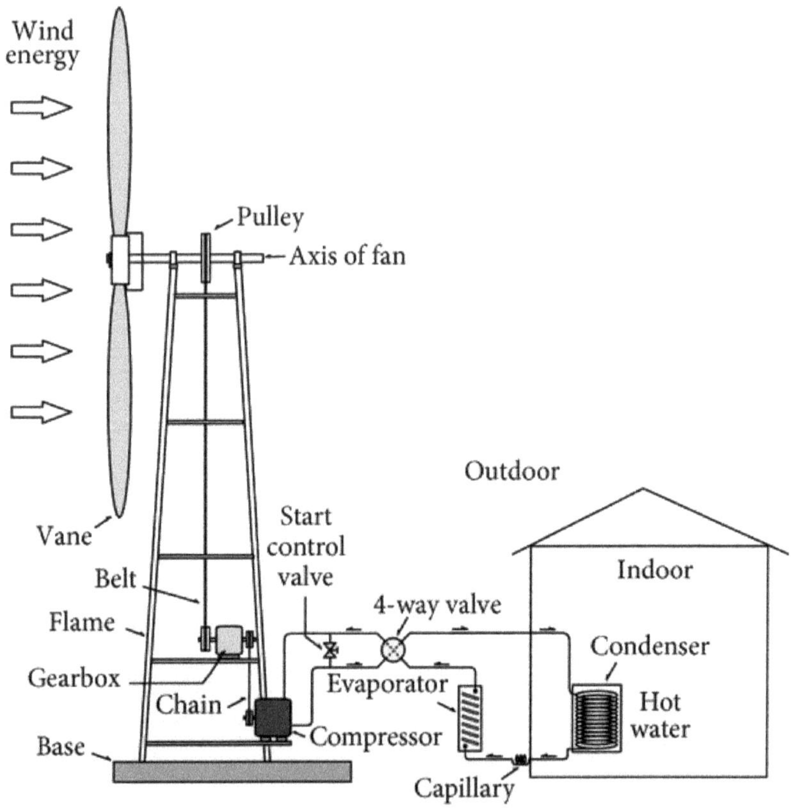

FIGURE 3: Winter model (heat pump) of wind directly forced heat pump.

11.4.1 THE WIND SYSTEM AND TRANSMISSION

The wind turbine consisted of three 1.6 m long fiberglass-reinforced polymers (FRP) vanes. The central axis is affixed onto the frame through the bearing that is installed with a pulley. A transmission is installed underneath the frame with a transmission gear ratio of 1 : 5. The ratio of diameter of the pulley installed on the central axis of wind turbine and that of the pulley of transmission is 5 : 1; therefore, the rotation speed of the wind turbine could be increased by 1 : 25 to drive the compressor of the heat pump system.

11.4.2 THE HEAT PUMP SYSTEM

The heat pump system as depicted in Figures 2 and 3 consisted of an open-type compressor, a condenser, a capillary, an evaporator, a start control valve, and a four-way valve. During the startup stage of the wind turbine, the control valve is open, under the connected discharge end and suction end of the compressor. The compressor starts up during the idle running, and the start control valve is slowly shut off when the rotation speed of the wind turbine reaches its maximal speed and the coolant starts to circulate in the cooling system. The cooling capacity and heating capacity of the heat pump in this experiment are 0.5 kW and 0.6 kW, respectively.

Two kinds of experimental modes are adopted in the experiment. Figure 2 shows the summer model, under which the indoor heat exchanger serves as the evaporator and the outdoor heat exchanger serves as the condenser. Figure 3 shows the winter model, under which the indoor heat exchanger serves as the condenser and the outdoor heat exchanger serves as the evaporator. The changeover between the two experimental models is made through operating the 4-way valve. In the summer model, the outdoor temperature of 35°C and the indoor temperature of 25°C; while in the winter model, the outdoor temperature and indoor temperature both are 10°C.

The experimental device measure system is as shown in Figure 4, where P1 is the suction pressure of the compressor; P2 is the discharge pressure of the compressor; T1 is the suction temperature of the compressor; T2 is the discharge temperature of the compressor; T3 is the outlet temperature of the condenser; T4 is the inlet temperature of the evaporator. The above measuring points are connected with the data logger, and the experiments are performed at the different average wind speeds of 3 m/s, 4 m/s, 5 m/s, and 6 m/s, respectively. Here accurate data could be obtained from the computer and be used as the basis for analysis and discussion.

11.5 RESULTS AND DISCUSSION

This section summarizes the results obtained performance of the proposed wind directly forced heat pump. The objective is to evaluate the working efficiency and the generalization capability of the wind directly forced heat pump. Summer model (chiller) and winter model (heat pump) of wind directly forced heat pump were examined during 60 minutes by working efficiency at the different average wind speeds of 3 m/s, 4 m/s, 5 m/s, and 6 m/s, respectively.

11.5.1 WORKING EFFICIENCY IN SUMMER MODEL (CHILLER) OF WIND DIRECTLY FORCED HEAT PUMP

Figure 2 shows the cool water producing model of the heat pump system driven directly by wind turbine. Figure 5 shows the measurement of the cool water producing capacity of the heat pump system driven directly by the wind turbine. In the experiment of the chiller, the outdoor temperature is 35°C and the indoor temperature is 25°C. The cooling capacity in the summer is measured with a 28 liters water storage tank, and the output capacity of the wind turbine is calculated to obtain the ratio between the cooling capacity of the heat pump system and the output capacity of the wind turbine. The working efficiency of the cooling of the wind turbine is also obtained. Figure 5(a) shows the data at the average wind speed of 3 m/s with the relative data calculated as follows.

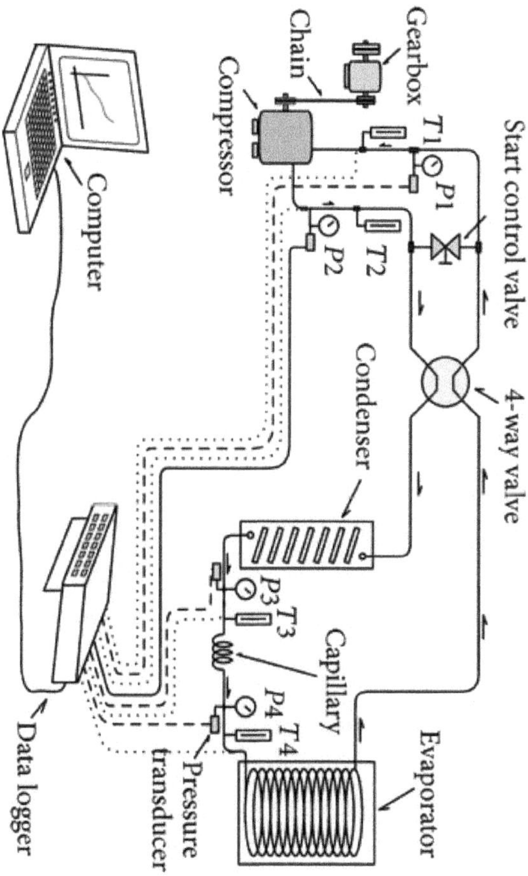

FIGURE 4: The experimental device and measure system.

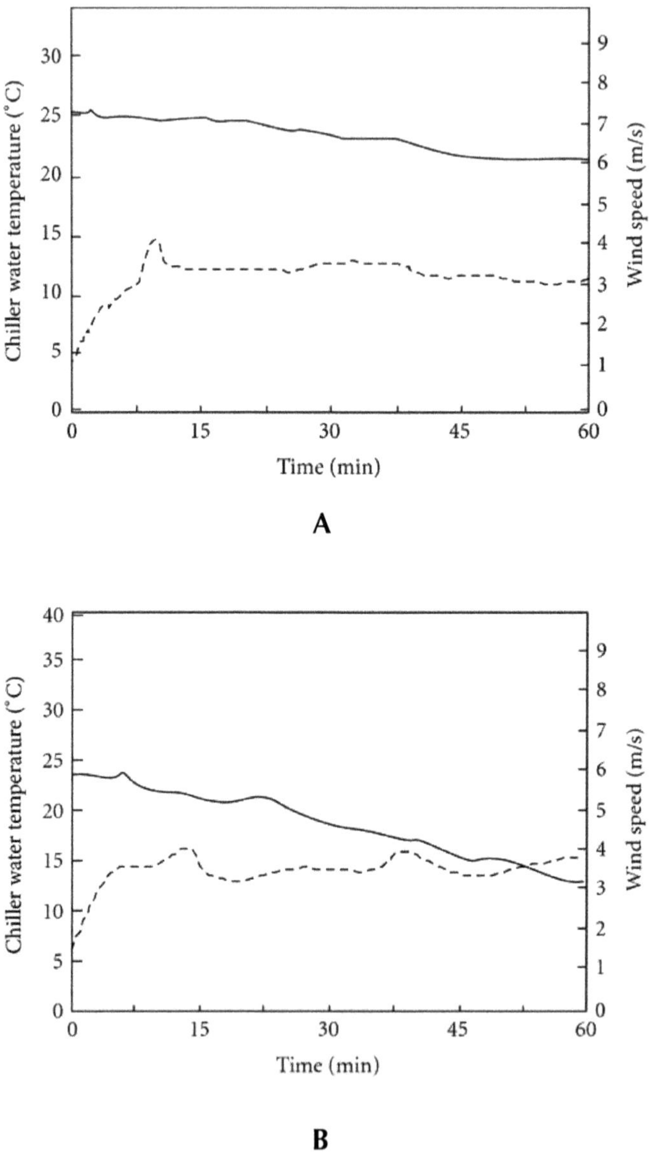

FIGURE 5: Experiment data of summer model (chiller operation) of heat pump.

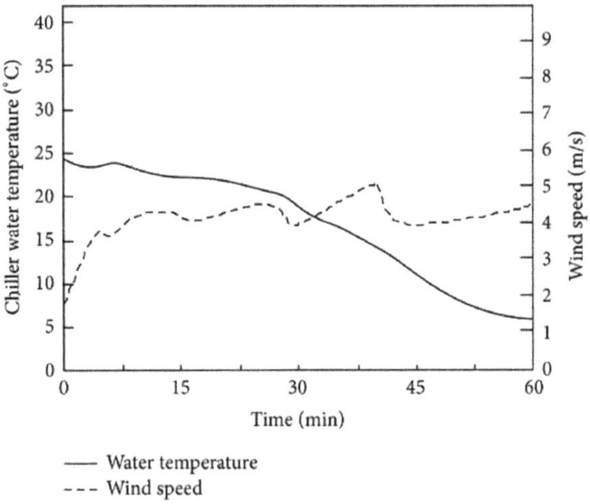

C

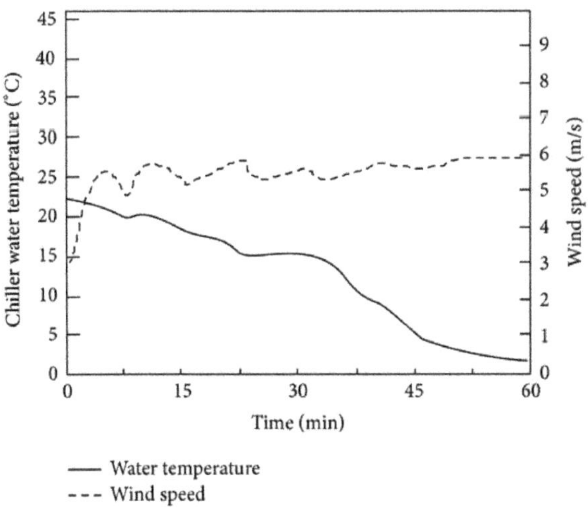

D

FIGURE 5: *Cont.*

Q_c is the cool water producing capacity of the heat pump system.

$$Q_c = m \times s \times \Delta t_c = 28 \times 4.186 \times 2 = 234.42 \text{ kJ} \quad (7)$$

E is the output capacity of the wind turbine.

$$E = 0.5\rho AV^3t = 478.57 \text{ kJ} \quad (8)$$

η_c is the efficiency of cooling of the wind turbine.

$$\eta_c = (Q_c/E) \times 100\% = 48.99\% \quad (9)$$

TABLE 1: Summation of the experimental condition for the summer model (chiller operation).

Experiment no.	1	2	3	4
Average wind speed (m/s)	3	4	5	6
Total duration (min)	60	60	60	60
Temperature change (°C)	25 ~ 23	24 ~ 19	25 ~ 14.5	23 ~ 3
Cooling energy, Q_c (kJ)	234.42	586.04	1230.68	2344.16
Wind energy, E (kJ)	478.57	1134.38	2215.58	3823.53
Working efficiency,	48.99	51.66	55.55	61.30
Climate condition	the outdoor temperature is 35°C and the indoor temperature is 25°C			

Figures 5(b)–5(d) indicate the wind turbine cooling data at the other three different wind speeds. Table 1 shows the comparison of experimental data of four different kinds of average wind speeds in Figure 5. The results reveal that when the wind speed increases, the energy produced by the wind turbine increases significantly. Thus, the cooling capacity increases, and the efficiency of the wind turbine cooling increases accordingly, as depicted in Figure 6.

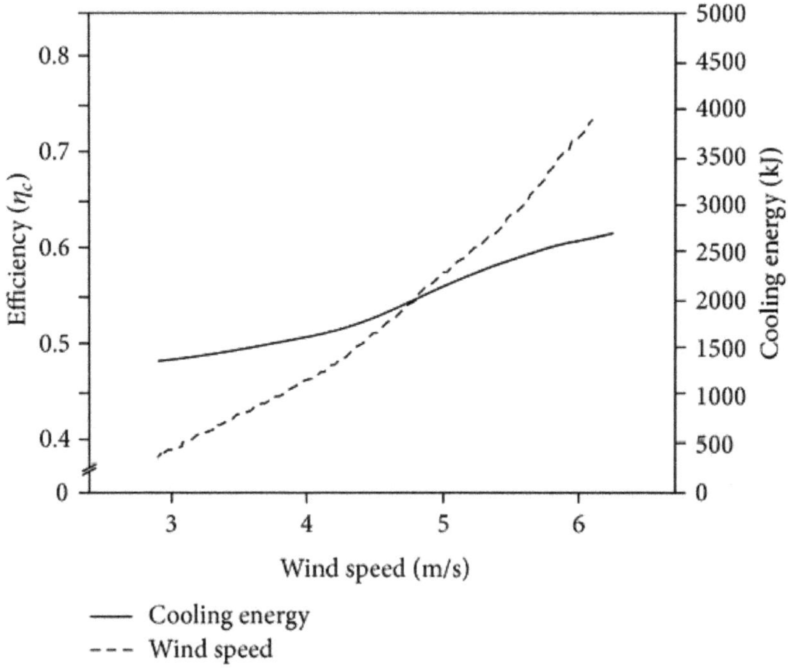

FIGURE 6: The relationship of efficiency and cooling energy versus wind speed for the summer model (chiller operation).

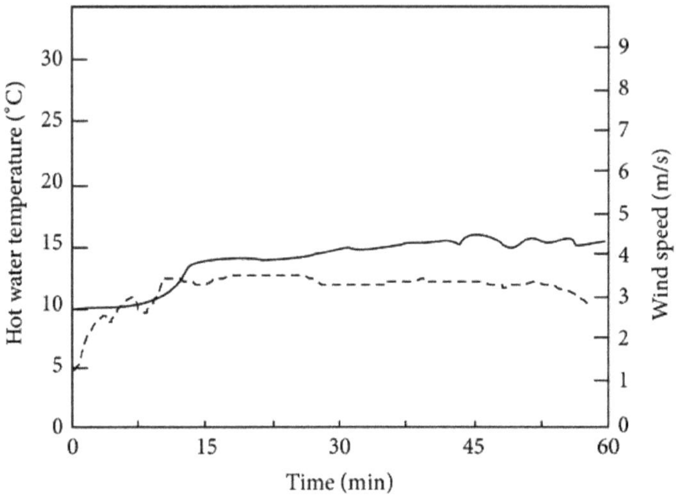

A

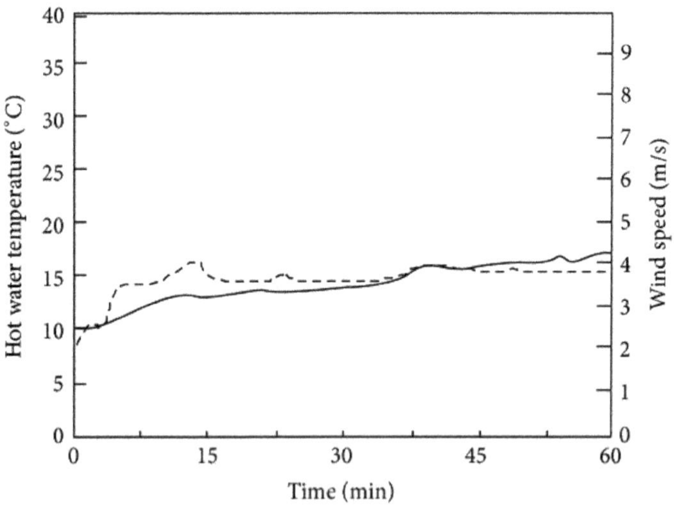

B

FIGURE 7: Experiment data of winter model (heat pump operation) of heat pump.

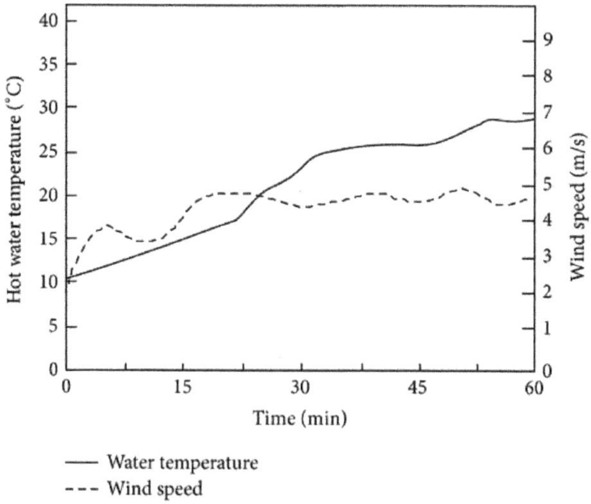

C

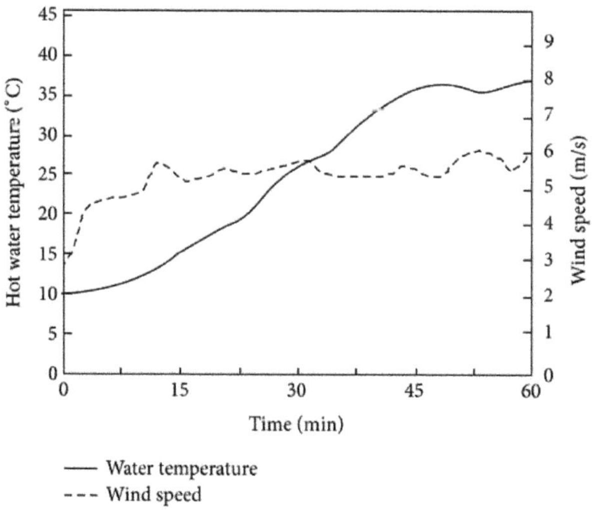

D

FIGURE 7: *Cont.*

11.5.2 *EFFICIENCY IN WINTER MODEL (HEAT PUMP) OF WIND DIRECTLY FORCED HEAT PUMP*

In order to change the heat pump system to the hot air producing model, only the direction of the 4-way valve needs to be adjusted, as shown in Figure 3. Figure 7 shows the measurement of the hot air producing efficiency of the heat pump system driven directly by the wind turbine. In the experiments of heat pump, the outdoor temperature and the indoor temperature both are 10°C. The heating capacity in the winter is measured with a 28 liters water storage tank, and the output capacity of the wind turbine is calculated to obtain the ratio between the heating capacity of the heat pump system and the output capacity of the wind turbine. The working efficiency of the heating of the wind turbine is also obtained. Figure 7(a) shows the data at the average wind speed of 3 m/s with the relative data calculated as follows.

Q_h is the hot air producing capacity of the heat pump system.

$$Q_h = m \times s \times \Delta t_h = 28 \times 4.186 \times 2.5 = 293.02 \text{ kJ} \quad (10)$$

E is the output capacity of the wind turbine.

$$E = 0.5\rho AV^3t = 478.57 \text{ kJ} \quad (11)$$

η_h is the efficiency of heating of the wind turbine.

$$\eta_h = (Q_h/E) \times 100\% = 61.23\% \quad (12)$$

Figures 7(b)–7(d) indicate the wind turbine heating data at the other three different wind speeds. Table 2 shows the comparison of experimental data of four different kinds of wind speeds in Figure 7. The results reveal that when the wind speed increases, the energy produced by the wind turbine increases significantly. However, the efficiency of the wind turbine heating in Figure 8 decreases. This phenomenon is because the heating is hard to be achieved when the outdoor temperature is only 10°C, so that the evaporator and outdoor air exchange thermal energy slowly. If heating capacity increases, it means more power will be consumed, and thus the working efficiency goes down consequently.

TABLE 2: Summation of the experimental condition for the winter model (heat pump operation).

Experiment no.	1	2	3	4
Average wind speed (m/s)	3	4	5	6
Total duration (min)	60	60	60	60
Temperature change (°C)	10 ~ 12.5	11 ~ 16.5	10 ~ 18.5	10 ~ 25
Cooling energy, Q_h (kJ)	293.02	644.64	996.27	1758.12
Wind energy, E (kJ)	478.57	1134.38	2215.58	3823.53
Working efficiency,	61.23	56.83	44.96	45.98
Climate condition	the outdoor temperature and the indoor temperature both are 10°C			

11.5.3 SUMMARY OF EFFICIENCY OF WIND DIRECTLY FORCED HEAT PUMP

Tables 1 and 2 summarize the working efficiency of wind turbine cooling at four different wind speeds and the working efficiency of wind turbine heating at four different average wind speeds, namely, 3 m/s, 4 m/s, 5 m/s, and 6 m/s. In the summer model (chiller), the outdoor temperature is 35°C and the indoor temperature is 25°C; the working efficiency of the cooling process is 48.99%, 51.66%, 55.55%, and 61.30%, respectively, whereas the average cooling working efficiency is 54.38%. In the winter model (heat pump), the outdoor temperature and the indoor temperature both are 10°C; the working efficiency of the heating process is 61.23%, 56.83%, 44.96%, and 45.98%, respectively, whereas the average heating working efficiency is 52.25%. The results verify that the heating efficiency is lower than the cooling efficiency. If other heat sources like underground heat source, solar energy, and other waste heat are available, then the heat pump heating efficiency could be further improved.

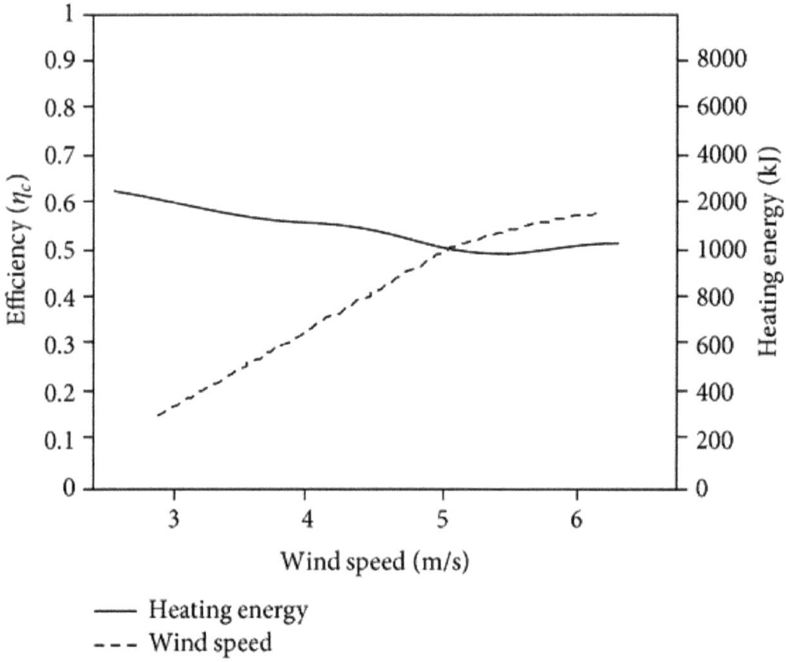

FIGURE 8: The relationship of efficiency and cooling energy versus wind speed for the winter model (heat pump operation).

11.6 CONCLUSIONS

The experiment results suggested that driving the heat pump system directly with the wind force can reduce the energy conversion losses during the processes of wind force energy converting to electric energy and electric energy converting to kinetic energy. Taking the air-conditioner load out of the electric load could reduce the equipment capacity of the power distribution project. The experimental results verified that the cool water producing efficiency of the heat pump system driven directly by wind force is 54.38% in the outdoor temperature of 35°C and the indoor temperature of 25°C; and the hot water producing efficiency is 52.25% in the outdoor temperature and indoor temperature both of 10°C. The proposed method compared with traditional method which theoretical efficiency is 42.19% can significantly improve the efficiency by over 10% in both cooling and heating processes. By the time this technology is maturely developed, it could be promoted to the areas with a large quantity of wind farms, such as islands, coast, desert, and high-rise buildings to achieve the purpose of energy conservation and carbon reduction.

REFERENCES

1. H. Holttinen, "Optimal electricity market for wind power," Energy Policy, vol. 33, no. 16, pp. 2052–2063, 2005.
2. M. Esteban and D. Leary, "Current developments and future prospects of offshore wind and ocean energy," Applied Energy, vol. 90, pp. 128–136, 2012.
3. R. W. Y. Habash, V. Groza, Y. Yang, C. Blouin, and P. Guillemette, "Performance of a contrarotating small wind energy converter," ISRN Mechanical Engineering, vol. 2011, Article ID 828739, 10 pages, 2011.
4. Y. Bai, T. T. Chow, C. Ménézo, and P. Dupeyrat, "Analysis of a hybrid PV/Thermal solar-assisted heat pump system for sports center water heating application," International Journal of Photoenergy, vol. 2012, Article ID 265838, 13 pages, 2012.
5. L. Q. Liu, Z. X. Wang, H. Q. Zhang, and Y. C. Xue, "Solar energy development in China—a review," Renewable and Sustainable Energy Reviews, vol. 14, no. 1, pp. 301–311, 2010.

6. J. B. Welch and A. Venkateswaran, "The dual sustainability of wind energy," Renewable and Sustainable Energy Reviews, vol. 13, no. 5, pp. 1121–1126, 2009.
7. I. Argatov and R. Silvennoinen, "Energy conversion efficiency of the pumping kite wind generator," Renewable Energy, vol. 35, no. 5, pp. 1052–1060, 2010.
8. P. Flores, A. Tapia, and G. Tapia, "Application of a control algorithm for wind speed prediction and active power generation," Renewable Energy, vol. 30, no. 4, pp. 523–536, 2005.
9. C. J. Lin, O. S. Yu, C. L. Chang, Y. H. Liu, Y. F. Chuang, and Y. L. Lin, "Challenges of wind farms connection to future power systems in Taiwan," Renewable Energy, vol. 34, no. 8, pp. 1926–1930, 2009.
10. T. J. Chang, Y. T. Wu, H. Y. Hsu, C. R. Chu, and C. M. Liao, "Assessment of wind characteristics and wind turbine characteristics in Taiwan," Renewable Energy, vol. 28, no. 6, pp. 851–871, 2003.
11. J. F. Manwell, J. G. McGowan, and A. L. Rogers, Wind Energy Explained: Theory, Design and Application, John Wiley & Sons, New Jersey, NJ, USA, 2002.
12. C. C. Ting, J. N. Lee, and C. H. Shen, "Development of a wind forced chiller and its efficiency analysis," Applied Energy, vol. 85, no. 12, pp. 1190–1197, 2008.

CHAPTER 12

The Feasibility of Wind and Solar Energy Application for Oil and Gas Offshore Platform

Y. K.TIONG, M. A. ZAHARI, S.F. WONG, AND S. S. DOL

12.1 INTRODUCTION

Nowadays, most of the offshore platforms are operated by burning of fossil fuels to generate electricity. The gas turbines and diesel generators on most platforms are used to drive pumps and compressors on board, powered up by combustion of conventional fuels. Continuous burning of these conventional fuels will generate about 80% of the total CO_2 and NOx emissions from offshore installations [1]. Thus, offshore platform are facing difficulties in term of operating their activities in an environmentally manner.

FIGURE 1: Wind turbines and solar panels on offshore platform [2]

Currently, great efforts have been taken in greening the energy sector by shifting the usage of fossil fuels to renewable energy in order to minimize current rate of fossil fuel usage and their ensuing effects of climatic changes. Renewable energy such as wind energy, solar energy and ocean energy are highly focused in the feasibility test of replacing conventional fuels to operate an offshore oil and gas platform. Some of the offshore platform has already started to exploit a renewable energy sources to generate power supply to lessen consumption of fuels. For example, a new offshore platform in the Southern North Sea as shown in figure 1 is operated by using their own energy generated through solar panels and wind turbines [2].

Wind and solar seems like ideal alternative sources of energy as it can provide an infinite amount of clean energy for offshore application. This paper will focus on the capability of renewable energy available around the offshore platform in order to determine the possibility of such utilization to meet the demand needed for these platform activities. Out of all renewable energies available these days, wind energy and solar energy are highly focused in this feasibility test since data collected predicted a rather promising result. Electrical power consumption of a platform is within the range from 10 MW to 50 MW, while the power consumption for smaller unmanned varies from 6 MW to 7 MW [1]. The targeted platform for this study is carried out at SHELL Malaysia Oil and Gas Sabah Water Platform, with the rated power consumption is approximately 10 MW. Based on the data collected from the targeted platform, the rated power output are determined in order to meet the demand of power consumption of this platform.

12.2 LITERATURE REVIEW

12.2.1 WIND ENERGY

In this feasibility study, SIEMENS SWT-4.0-120 is selected due to its suitable cut-in speed as low as 2.9 m/s. The wind turbine selected is a direct drive horizontal axis wind turbine, with the latest patented Quantum Blade technology used by Siemens, and able to generate rated power output as high as 4 MW with rated wind speed of 13 m/s [3].

In order to determine the rated power output, P_{rated} for specific month, equation (1) is used. Where, A_r is the swept area of the rotor; PD is the power density by the wind; η is the efficiency of the selected wind turbine stated by the manufacturer [4].

$$P_{rated} = A_r \times PD \times \eta \tag{1}$$

The swept area, A_r by the rotor of wind turbine can be determined by using equation (2) where, D is the diameter of the rotor for the selected wind turbine.

$$A_r = \pi \times (D/2)^2 \tag{2}$$

The power density, PD of the wind is determined by the equation (3) where, ρ_{air} is the air density at 50 m above sea level and v is the wind speed.

$$PD = 0.5 \times \rho_{air} \times (v^3/1000) \tag{3}$$

For Siemens SWT-4.0-120, the efficiency for the selected wind turbine is 32%, giving the value of η is equal to 0.32 [3].

12.2.2 SOLAR ENERGY

For the solar energy, the selected photovoltaic panel used for this feasibility test is LG290 N1C-G3 MonoX, with the panel area of 1.64 m^2 and peak power output of 290 W [5]. The power output for photovoltaic panel can be determined via the equation (4):

$$P = A_p \times r \times H \times PR \tag{4}$$

where, A_p = area of solar panel in m^2, r = solar panel yield (%), H = average solar radiation on panels (W/m^2) and PR = Performance ratio.

Previous studies suggested that the performance ratio is set as 0.75 as default when solar panel is tilted at 45° of the direction of irradiation [6]. This value varies depending on the shadings present around the solar panel, as well as the cleanliness of the solar panel itself which is referring to the dust accumulation on solar panel.

Solar panel yield, r can be determined by the equation (5) where, PE is the electrical power of solar module selected; A_p is the area of the solar module.

$$\text{Solar panel yield, r (\%)} = PE/10A_p \tag{5}$$

12.3 DATA

12.3.1 WIND ENERGY

In this research, wind speed data surrounding the SHELL Sabah Water Platform is provided by Sarawak SHELL Bhd. With the tabulated data, the average monthly wind speed hitting around the platform 50 m above sea level are plotted as in figure 2.

Based on figure 2, the wind speeds around the SHELL Sabah Water Platform are within a range of 3.22 m/s to 6.07 m/s. A sample calculation to determine the rated power output is shown using the average wind speed on January (6.07 m/s). The wind turbine specification and wind parameter are listed in table 1.

The power output for January is calculated using the equations discussed before. The swept area by the rotor of wind turbine is given by equation (2):

$$A_r = \pi \times (D/2)^2 = \pi \times (120/2)^2 = 11{,}310\ m^2 \tag{6}$$

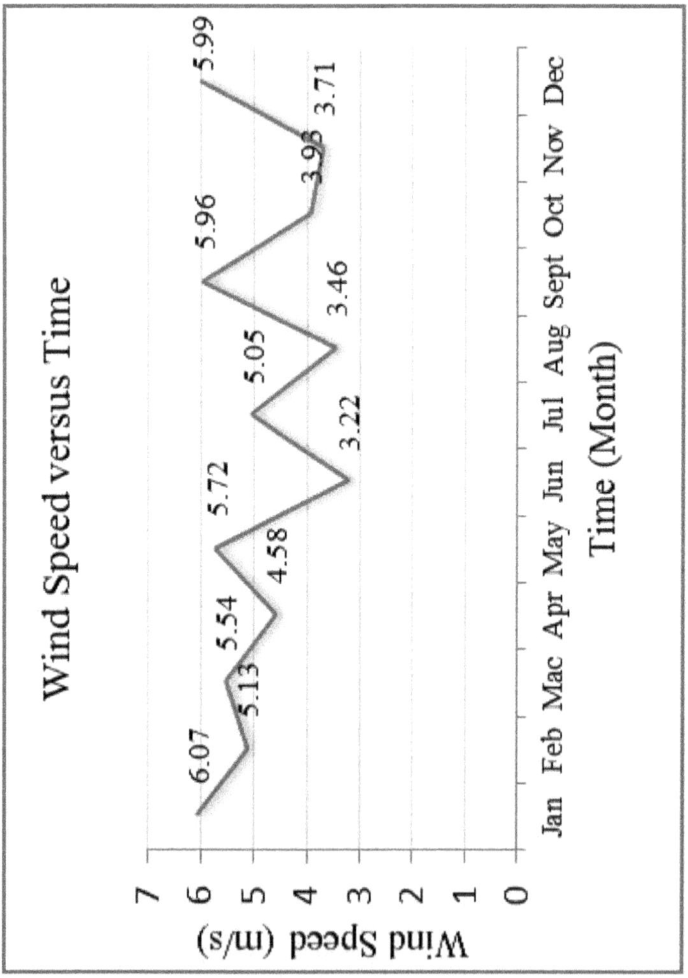

FIGURE 2: Average Wind Speed collected on SHELL Sabah Water Platform in 2008

TABLE 1: Wind turbine and wind parameters

Wind Turbines (SIEMENS SWT-4.0-120)	
Rotor diameter, D [m]	120
Blade length, L [m]	58.5
Efficiency, η [%]	32
Wind	
Air density, ρ_{air} [kg/m^3]	1.225 at 50 m above sea level

The power density, PD by the wind onto the wind turbine is determined by the equation (3):

$$PD = 0.5 \times \rho \times v^2 = 0.5 \times 1.225 \times 6.07^2 = 137 \text{ kg/s}^2 \quad (7)$$

Then, the power output for specific wind turbine for January can be determined via the equation (1):

$$P_{rated} = A_r \times PD \times \eta = 11310 \times 137 \times 0.32 = 495{,}830 \text{ W} = 495 \text{ kW} \quad (8)$$

12.3.2 SOLAR ENERGY

The solar irradiation available in the surrounding of SHELL Sabah Water Platform is provided by Sarawak SHELL Bhd. With the tabulated data, the average monthly solar irradiation hitting around the platform are plotted as in figure 3.

However, as the energy based on solar is dependent on the availability of the sun, additional graph is plotted based on daily solar irradiation to determine the peak hour of solar irradiation. With the collected data, the average daily solar irradiations hitting around the platform on are plotted as in figure 4.

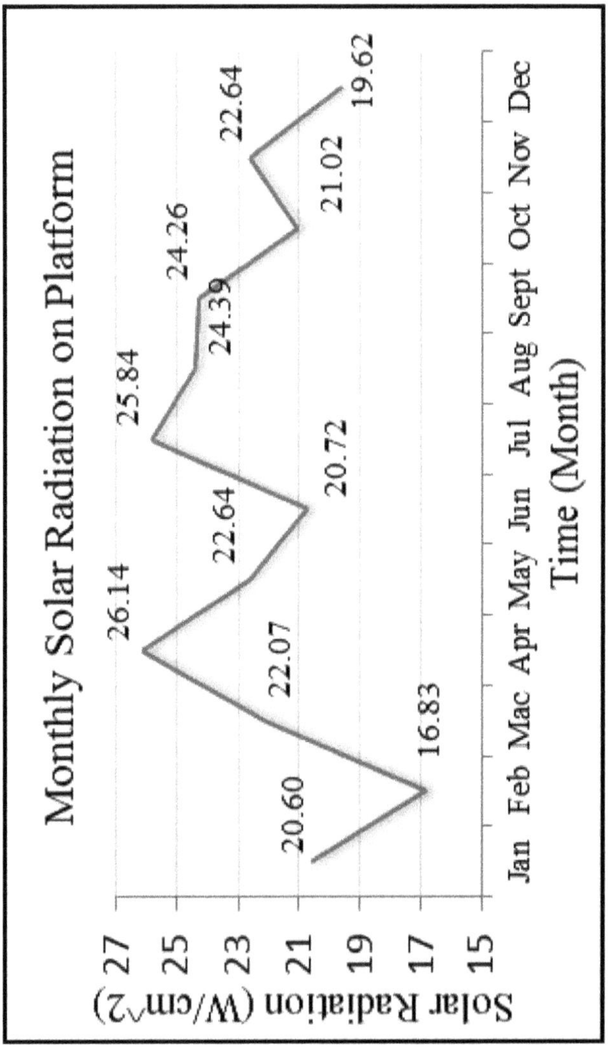

FIGURE 3: Average monthly solar radiation on SHELL Sabah water platform in 2008

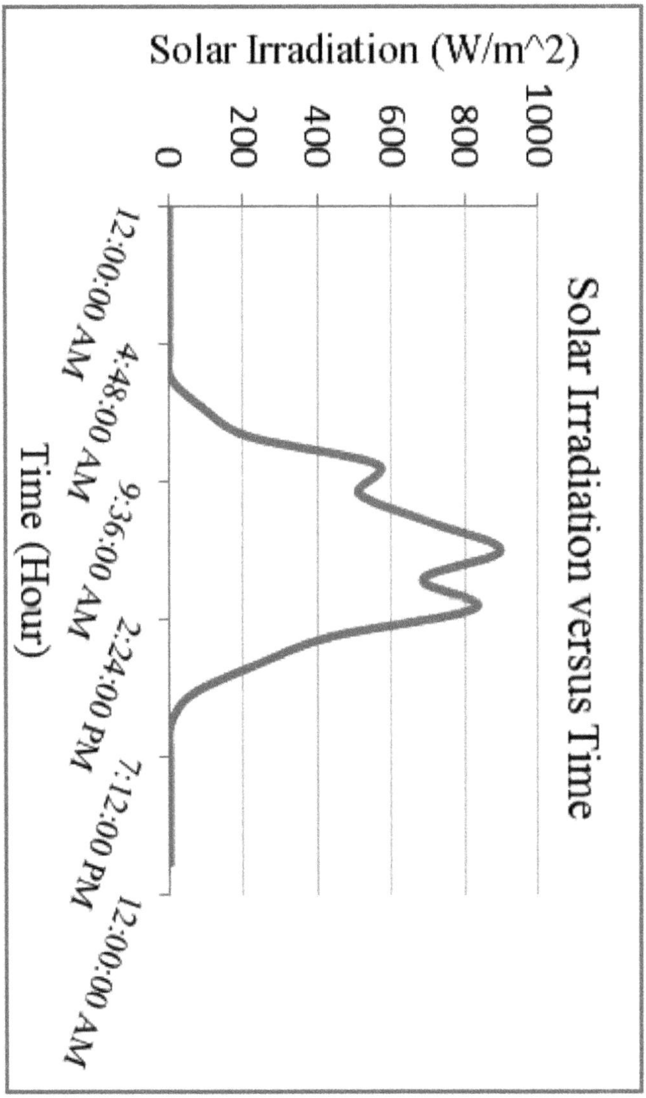

FIGURE 4: Daily solar irradiation peak hour on SHELL Sabah water platform.

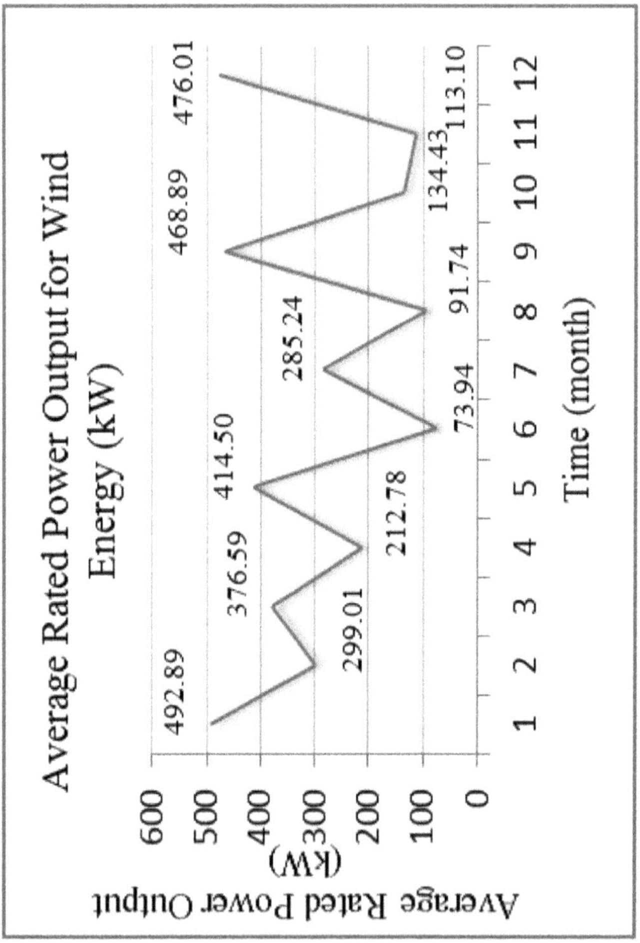

FIGURE 5: Monthly average rated power output for SIEMENS SWT4.0-120

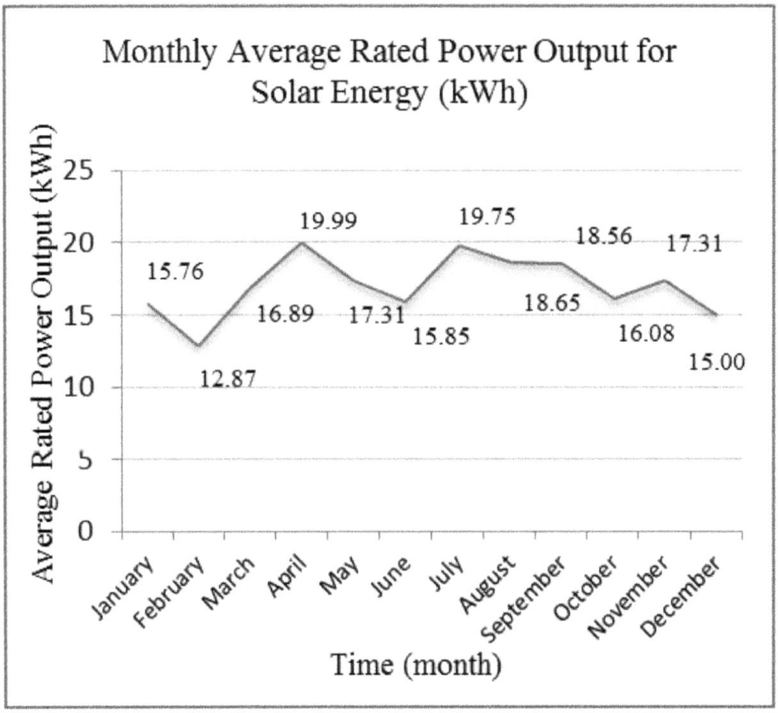

FIGURE 6: Monthly average rated power output for LG290 NIC-G3 MonoX

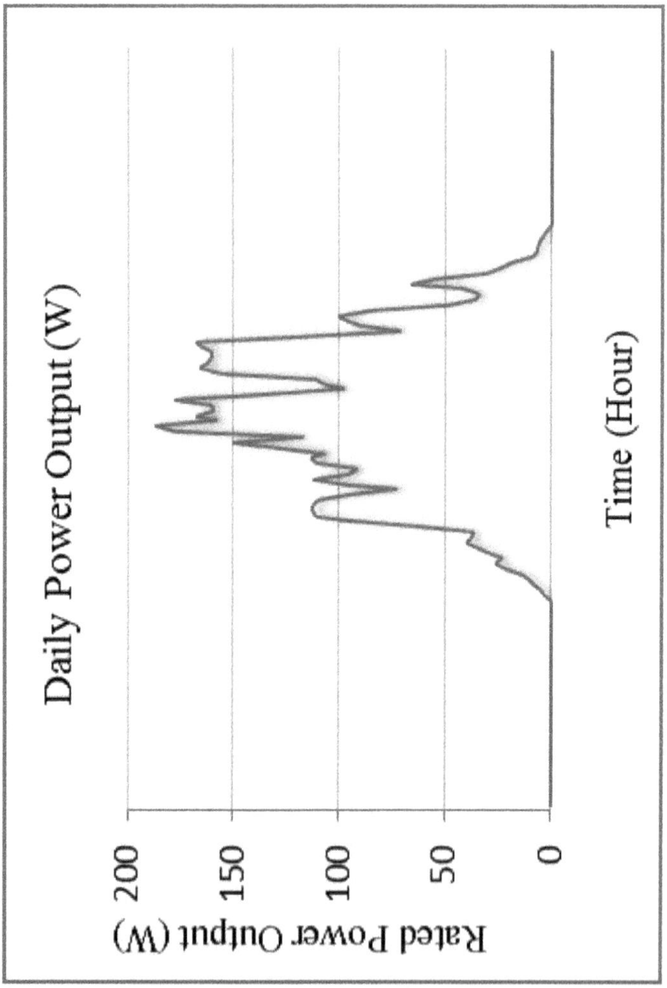

FIGURE 7: Daily average rated power output for LG290 NIC-G3 Monox

FIGURE 8: Multiple units of wind turbines and solar panels installed on offshore platform (conceptual design from [8])

Based on figure 3 and 4, the solar irradiation on the SHELL Sabah water platform varies within a range of 16.83 W/cm^2 to 26.14 W/cm^2 with the peak hour starting from 6 am to 6 pm. A sample calculation to determine the rated power output is shown using the average solar irradiation collected on April (26.14 W/cm^2). The selected photovoltaic panel used for this feasibility test is LG290 N1CG3 MonoX, with the panel area of 1.6 4m^2 and peak power output of 290 W [5]. While its performance ratio is set as default which is 0.75 based on the previous study [6]. The power output for April is calculated using the equations discussed before. Solar panel yield, r can be calculated using equation (5):

$$r\,(\%) = PE/10A_p = 290/10(1.64) = 17.68\% \qquad (9)$$

The average solar irradiation collected in April is converted from 26.14 W/cm^2 to 94.10 kWh/m^2. Then, the power output for photovoltaic panel can be calculated via the equation (4):

$$P_{rated} = A_p \times r \times H \times PR = 1.64 \times 0.1768 \times 94.1 \times 0.75 = 20 \text{ kWh} \qquad (10)$$

12.4 RESULT AND DISCUSSION

The rated power output calculated for each month based on selected wind turbine for wind energy application are plotted as shown in Figure 5.

Based on the figure 5, the monthly average rated power output generated by selected wind turbine SIEMENS SWT4.0-120 varies between the ranges of 73.94 kW to 492.89 kW. The highest power output was generated in January while the lowest month of power generated by this turbine was in Jun.

Based on the figure 6, the monthly average rated power output generated by solar panel LG290 NIC-G3 MonoX varies between the ranges of 12.87 kWh to 19.99 kWh. The poorest month of solar irradiation falls on

February with the power output generated 12.87 kWh while the highest month of solar irradiation falls in April with the power output generated 19.99 kWh.

Based on daily power output in figure 7, at time between 12:00 a.m. and 6:00 a.m. as well as between 6:00 p.m. up till the next morning, the solar irradiation recorded is zero, indicating that the system has its downtime during this period of time. This is due to the availability of the sun [7]. The peak hour of daily solar irradiation falls at time between 11 a.m. to 3 p.m.

Based on the results discussed before, these wind turbine and solar panel can be installed in multiple units to form an array system in order to provide sufficient power supply to the selected offshore platform (figure 8). In order to accommodate the rated power output of 10 MW, the solar irradiation investigated should be from the poorest out of all months and days of all data collected to form an array. Moreover, an integrated system should be considered using both of these wind turbine and solar panel applications [9]. This is due to the complementary power output generated by these two forms of energy under seasonal variant in weather conditions. The solar panel plays a major role in generate highest power output in the summer months when the output from the wind turbine is at the lowest and vice versa during the monsoon season.

12.5 CONCLUSION

The solar and wind energy application in generating energy is a viable solution of the current energy extraction problem for offshore application. In this paper, it showed that the highest power output generated for wind energy application is equal to 492 kW while for solar energy application power generated is equal to 20 kW. As a single unit of solar or wind turbine system is insufficient in order to provide power supply to offshore platform application, these applications can be installed in multiple units in order to generate sufficient power supply. Both of these systems can also be integrated to each other due to their different capabilities in generated power output based on weathers and climates condition.

REFERENCES

1. He W, Uhlen K, Hadiya M, Chen Z, Shi G and Rio E 2013 Case study of integrating an offshore wind farm with offshore oil and gas platforms and with onshore electrical grid Journal of Renewable Energy 1-10
2. Rosebro J 2006 Fossil-fuel platform runs on renewable energy http://www.greencar-congress.com/2006/04/fossilfuel_plat.html
3. Siemens 2014 Siemens 4.0MW offshore wind turbine http://www.energy.siemens.com/
4. NPower nd Wind turbine power calculations The Royal Academy of Engineering
5. Tecnospot 2013 MonoX Neon LG290NiC-G3 http://www.lg.com/uk/solar
6. Hanid M, Ramzan M, Rahman M, Khan M, Amin M and Amir M 2012 Studying power output of PV solar panels at different temperatures and tilt angles ISESCO Journal Science and Technology 8 9-12
7. Goffman E 2008 Why Not the Sun? Advantages of and Problems with Solar Energy Journal of ProQuest Discovery Guides
8. Tiong Y K 2013 The renewable energy application for oil and gas offshore platform Thesis (B.ME.) Curtin University Malaysia.
9. Zahari M A and Dol S S 2014 Application of vortex induced vibration energy generation technologies to the offshore oil and gas platform: The preliminary study International Journal of World Academy of Science, Engineering and Technology 8(7) 1331-34

CHAPTER 13

Economic Feasibility Analysis of the Application of Geothermal Energy Facilities to Public Building Structures

SANGYONG KIM, YOUNG JUN JANG, YOONSEOK SHIN, AND GWANG-HEE KIM

13.1 INTRODUCTION

Demand for improvements in quality of life, as well as for various benefits necessitates the provision of a consumer-centred construction environment, even within the building sector. To satisfy consumer demands related to indoor and outdoor environments, a construction environment system should accommodate both natural and artificial controls. To accomplish this end, an optimal building structure should be built in composite harmony with construction plans, the latter including the facility system plan, the spatial plan and the structural plan. Interest in reducing greenhouse gas emissions arising from energy demand within building structures has gradually been increasing, this being a component of optimal building structure; however, there are technical limitations to the extent to which energy consumption can be reduced simply by maximizing the efficiency of fossil energy in conventional use. One of the approaches to addressing this problem is to reduce fossil fuel consumption using new renewable energy resources [1]. To fundamentally increase the efficiency of energy consumption, there is thus an urgent need to develop new technologies that can utilize these eco-friendly renewable energy resources [2].

The EU stipulates that all new building structures to be built from 2019 onwards must produce more energy than they consume; similarly, with a view toward achieving zero-energy buildings by 2025, compulsory regulations and obligatory expansion plans have been established in the U.S. [3]. Other advanced countries have likewise made efforts to expand new renewable energy facilities and increase the supply ratio. In line with this trend, Korea has also made efforts to increase the supply ratio of new renewable energy from the fairly low level of 1.6% in 2012. For these reasons, planning for energy consumption capacity and savings should start at the construction project planning phase.

Geothermal power is cost effective, reliable, sustainable and environmentally friendly. It is also available 24 hours/day which can be used as the base load. Historically, it has been limited to areas near tectonic plate boundaries. However, recent technological advances have dramatically expanded the range and size of this source, especially for applications, such as home heating, opening a potential for widespread exploitation, as has been described in this study. In this study, the geothermal system, a renewable energy facility applied to multiple building structures, was employed to analyse relative reductions in energy consumption and energy use cost and to determine the energy savings cost related to the use of this system in consideration of the lifecycle cost (LCC), which includes initial investment costs, repair and replacement cycles for major materials. The ultimate aim is that of proposing an effective plan for geothermal system selection at the construction project planning stage through economic feasibility analysis.

13.2 METHODOLOGY

This study is limited to the application of new renewable energy to a building structure with geothermal facilities, based on statistics and on previous studies. The target buildings selected for this research were three public buildings, within which the installation of a new renewable energy facility is compulsory. In all cases, the renewable energy facility, which varies in size across the three buildings, had already been designed. By analysing the design data of the buildings, the volume of new renewable energy was converted into that of the energy applied to the conventional energy facility, and the buildings were modelled with no application of the new renewable energy source. Using total energy consumption evaluation program (ECO2-OD), the compulsory supply volume of new renewable energy was calculated, and geothermal facility installation sizes in the proportions of 100% and 25% of compulsory supply volume, respectively, were set and applied in each case.

Energy consumption and primary energy consumption of the building structures were calculated to perform a comparative analysis of changes related to the application of the geothermal energy system and to determine the changes in the energy volume used. A comparative analysis of the economic feasibility of geothermal energy application was conducted between buildings with geothermal facilities and those without, based on reductions in energy cost. The initial construction cost was applied, together with the cost of the redesigned specification for buildings with geothermal facilities. Where the volume of the conventional energy facility was replaceable, the volume change of the conventional facility resulted from the application of the geothermal facility, and the removal of the facility was reviewed and reflected in the initial investment cost. The useful life was set at 40 years in the LCC analysis; the analysis cycle was set at 10 years, and the repair and replacement cycles of major materials were chosen to ensure the reliability of maintenance cost calculations.

13.3 LITERATURE REVIEW

Many studies on the application of new renewable energy facilities to building structures have been conducted, through which plans have also been proposed to improve the reliability and economy of such applications. Rezaie et al. [4] divided case buildings by usage to analyse the economy, efficiency and energy emission of geothermal, solar-powered, photovoltaic and hybrid power systems. Visa et al. [5] studied the energy state required before and after the installation of a photovoltaic energy system in terms of efficiency and economic feasibility and performed an analysis of the latter. Cucchiella et al. [6] conducted a performance evaluation of a photovoltaic energy system installed in building structures to analyse the time required to retrieve investment cost and the extent of influence of climate and the energy consumption behaviour of residents within the area of installation. Francisco and Batlles [7] conducted a comparative analysis between a cooling system that applied photovoltaic energy and a conventional system and predicted the rate of reduction in energy consumption capacity as an alternative to reducing energy consumption. Similarly, many studies have been actively conducted to examine the utilization of new renewable energy resources in different countries in order to reduce energy consumption. However, there are no clear-cut criteria for the selection and application of such renewable energy facilities at the building structure planning phase, and it is difficult to select a facility that will achieve effective energy savings and secure economic feasibility. Furthermore, the determination of a new renewable energy system for building structures should occur at the planning and design phases, and its applicability and characteristics need to be taken into account, because the energy source may vary depending on the appearance and use of a building structure.

Previous studies performed in Korea can be subdivided into two categories—those examining the current state and application of new renewable energy within building structures and energy production volume and economic feasibility analyses of new renewable energy sources. Jung et al. [8] proposed a process to integrate a new renewable energy system into a construction design factor to enable the application of new renewable energy; the authors analysed systems according to their characteristics to apply and analyse applicable new renewable energy systems depending on the design process. Kang et al. [9] analysed the building energy substitution rate for public buildings in which 5% of the total construction cost was invested into the installation of a new renewable energy facility and presented a plan for efficiency improvement with no additional cost, using such a facility. As a component of the fundamental data to develop plan criteria for the application of a new renewable energy system to building structures, Kim et al. [10] proposed a direction for domestic construction planning based on an analysis of building structures that apply new renewable energy in Korea and Germany, focusing on photovoltaic and geothermal energy. Yoon et al. [11] suggested an option for the development of a new renewable energy planning tool that reflects various requirements, including the supply ratio of new renewable energy, the selection of a system type, the time of application and the method. Seo et al. [12] derived an optimal application plan by researching the current state of new renewable energy penetration and reviewing alternatives for an efficiency improvement plan according to changes in the volume of conventional new renewable energy systems. In addition, reviews of energy production volume and economic feasibility relating to new renewable energy resources applied to building structures have been actively conducted. Kim and Kim [13] carried out economic feasibility evaluations of photovoltaic, wind, small hydro and bio-gas power systems and presented a new renewable energy application plan. Kim et al. [14] performed an LCC analysis of the application of new renewable energy to reduce energy consumption and to analyse the energy-saving effect resulting from the reduction in the energy consumption cost and the retrieval period of the initial investment cost upon application of the new facility; the authors also proposed a scheme for economic building structure planning. Lee et al. [15] conducted an analysis of energy intensity by building use based on an analysis of the actual state of energy use in public office facilities, with the enforcement of the law compelling the installation of new renewable energy facilities within public organizations. However, assessments conducted in previous studies have addressed the future applicability of new renewable energy sources at the level of fundamental data by exploring the current state of new renewable energy use, the actual state of management and its problems and user satisfaction; the limitation of these studies thus lies in the lack of understanding of a direction to improve the management of a new renewable energy system. In addition, there are no concrete analysis processes and criteria for the analysis of energy production volume and economic feasibility, and a cyclical analysis of the entire lifecycle of building structures has not been performed, resulting in many limitations to actual applications of resolutions. The summary of previous studies is listed in Table 1.

TABLE 1: The summary of previous studies. LCC, lifecycle cost.

Authors	Year	Research Aim
Jung et al.	2008	Applying and analysis of a new renewable energy design process
Kim & Kim	2008	Presenting a new renewable energy application plan
Rezaie et al.	2011	Analysis of energy emission quantity based on renewable energy options
Kang et al.	2011	Analysis of renewable energy and building energy substitution rate
Kim et al.	2011	LCC analysis of the application of new renewable energy to reduce energy consumption
Cucchiella et al.	2012	Performance evaluations of integrated photovoltaic systems
Lee et al.	2012	Analysis of energy intensity by building use based on the actual state of energy use
Kim et al.	2012	Applying new renewable energy focusing on photovoltaic and geothermal energy
Francisco & Batlles	2013	Forecasting energy savings rate by applying solar energy
Yoon et al.	2013	Suggesting an option for the development of a new renewable energy planning tool
Seo et al.	2013	Improvement plan according to changes in volume of conventional new renewable energy
Visa et al.	2014	Economic analysis of the renewable energy mix in a building

13.4 THE GEOTHERMAL SYSTEM

A geothermal system uses geothermal heat to achieve an increase in temperature, with the latter maintained at a certain level deeper than 15 m below the surface of the Earth, regardless of atmospheric temperature. Geothermal energy can be largely subdivided into geothermal and land surface heat. Geothermal heat is heat energy continuously generated by the decay of radioactive isotopes in the core of the Earth, i.e., the energy that magma emits toward the surface of the Earth. Land surface heat can be classified as shallow or deep geothermal heat depending on depth below the surface of the Earth [16]. Shallow geothermal heat averages 5–20 °C at less than 15 m from the surface of the Earth. To use shallow geothermal heat, a 100- to 300-m-deep borehole is drilled to lay a geothermal heat exchanger, which provides energy to a building structure for heating and cooling using a heat pump and an air handling unit, as shown in Figure 1. In the case of deep geothermal heat, the Earth is excavated to 3 km below the ground to obtain steam with a temperature of 65 °C or higher to generate electricity through a steam turbine [17]. A geothermal system for cooling in summer operates as follows: the high temperature and high pressure cooling gas compressed by the compressor of a heat pump exchanges heat at the geothermal heat exchanger, with conversion into a moderate temperature and high pressure liquid, resulting in cooling effects when the liquid evaporates in the indoor evaporator at low temperature and in a low pressure state through expansion in thc expander. On the other hand, in winter, the geothermal heat pump system works in the opposite direction, absorbing geothermal heat and supplying it to the inside of a building [18].

Geothermal systems can be broadly categorized as either open or closed, depending on the circuit composition of the heat exchanger that collects the geothermal heat. An open loop system is one in which the pipe that carries water supplied from phreatic and underground sources is applied to locations within a basin with effluent water. A closed loop system is one in which water circulates within the pipe to collect (exchange) geothermal heat. Closed loop systems are classified as vertical or horizontal based on the loop type, as shown in Figure 2. The vertical type ranges from 100–150 m below ground, while the horizontal type ranges from 1.2–1.8 m below ground.

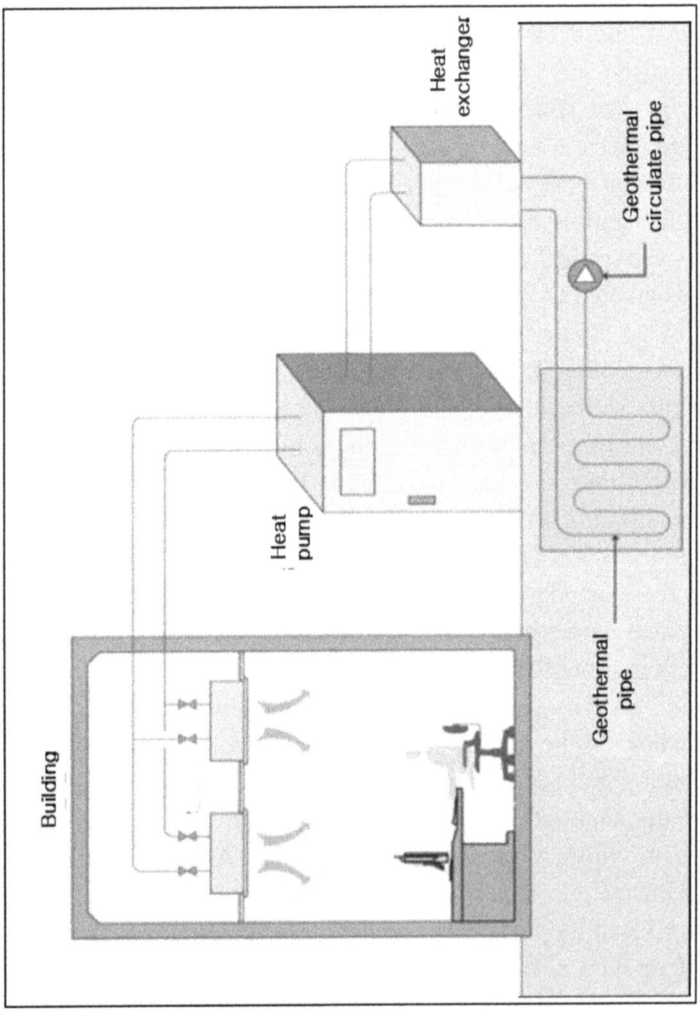

FIGURE 1: The structure of a geothermal system.

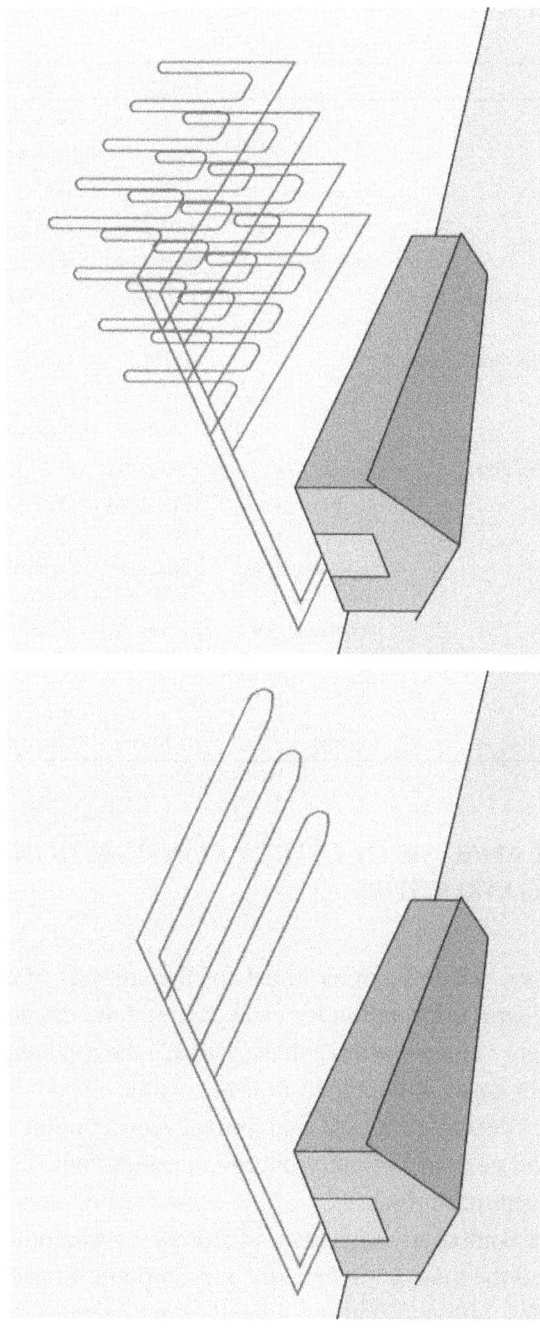

FIGURE 2: Vertical (left) and horizontal (right) types of closed loop systems.

TABLE 2: The description of an electric heat pump (EHP).

Description		Case 1	Case 2	Case 3
Type of cases			Public building	
Structure			Reinforced concrete	
Size of building	No. of floors		Basement floor = 1, Ground floors = 3	
	Building areas	774.42 m^2	1495 m^2	2786.57 m^2
	Total areas	1997.83 m^2	4006.58 m^2	12,690.79 m^2
Heating and air-conditioning areas		1421.58 m^2	2699.4 m^2	7292.16 m^2
Outer wall area/window area		20.45%	18.81%	32.67%
Type of cases			Public building	
Structure			Reinforced concrete	
Main energy facilities	Heating (kW)	EHP: 339.6	EHP: 420.1	EHP: 2358
	Air-conditioning	EHP: 301.6 kW	EHP: 377 kW	EHP: 2289.6 kW
	Hot-water supply (kW)	Gas boiler: 41.86	Lamp oil boiler: 116.28	Gas boiler: 244 kW × 2
	Ventilation (kW)	Exhaust fan lamp: 6.706	Exhaust fan lamp: 5.47	Air-conditioning equipment lamp: 35.72
	Lighting	6.64 W/m^2	8.02 W/m^2	8.00 W/m^2

13.5 CASE ANALYSIS OF ENERGY CONSUMPTION CAPACITY BY BUILDING STRUCTURE

The following are to be determined for the analysis of energy consumption: (1) general information for each case and the major energy load factors for energy consumption calculation; (2) the total annual energy consumption for cases in which no new renewable energy facility is applied, the energy consumption cost and annual consumption by unit area, the consumption cost and primary energy consumption; (3) the total annual energy consumption for cases where a geothermal energy facility is applied; (4) a comparative analysis of energy consumption and consumption cost and the total annual energy consumption for each case, considering examples with and without a geothermal energy facility; and (5) the energy-saving effects and cost savings resulting from the application of a geothermal energy facility for all cases and the energy consumption and energy consumption cost by the scale of the building structure following the application of the geothermal facility.

As shown in Table 2, three public office buildings of a similar size, all of which are subject to compulsory installation of a new renewable energy facility, were selected as the target cases to be analysed. The location of all cases is a central city of South Korea. Energy consumption and energy consumption costs were analysed for each building; subsequently, the same parameters were measured following the application of a geothermal energy system, for use in calculating and analysing the LCC of the buildings.

13.6 ANALYSIS OF ENERGY CONSUMPTION AND COST

Table 3 shows the total annual energy consumption and cost for each case, calculated using ECO2-OD. Total annual energy consumption and cost, which were compared across cases following the application of the geothermal energy facility, were calculated based on heating, cooling and hot-water supply calculated in consideration of cooling and heating areas and lighting and ventilation calculated in consideration of gross floor area and the sum total. In terms of consumption cost, the basic rate for contract electricity capacity applied to each building is fixed, and this was not included in the analysis of consumption cost. In the analysis of total annual energy consumption and consumption cost, the area to which individual energy load is applied was calculated by multiplying each cooling and heating area, while the lighting and ventilation load were calculated by multiplying gross floor area. There were therefore differences in total energy consumption capacity and consumption cost. For this reason, it was hard to set evaluation criteria for the comparison of each case, and this data was not utilized for comparisons; this aspect would, however, comprise a significant calculation variable for the application of a new renewable energy facility.

The cost of the LCC of CO_2 ($LCCO_2$) emission can be estimated by multiplying the required material cost of a component with the CO_2 emission basic units of the identified major construction materials. The multiplied cost should be converted into the current market-traded CO_2 emission price. As mentioned earlier, there are various markets for trading emission rights with the intention of controlling air pollution in developed countries. Among various markets, the price of the EU Allowance (EUA), which is traded in the European telecommunication standards, is adopted to calculate the $LCCO_2$, and the average price of CO_2 emissions (from 2005 to 2009), as suggested by European climate exchange, is applied. The average price is 19.73 EURO/ton. In addition, the average Euro:Won exchange rate in 2009 is applied, which is the standard currency in the ECX.

TABLE 3: Annual energy demand quantity and use of cases.

Description		Case 1	Case 2	Case 3
Energy demand quantity (kWh/year)	Heating	68,981.18	118,053.14	365,861.47
	Air-conditioning	26,977.01	60,118.07	201,622.22
	Hot-water supply	18,091.80	35,867.80	66,520.70
	Lighting	31,160.87	75,762.22	238,485.33
	Ventilation	31,155.86	12,350.45	145,950.22
	Total	176,366.72	302,151.68	1,018,439.93
Energy use cost (Won/year)	Heating	6,044,821	11,415,739	35,378,804
	Air-conditioning	2,363,995	5,813,417	19,496,869
	Hot-water supply	1,440,518	4,040,650	5,264,658
	Lighting	2,730,627	7,326,207	23,061,531
	Ventilation	2,730,188	1,194,289	14,113,386
	Total	15,310,149	29,790,302	97,315,248

13.7 ANALYSIS OF ENERGY CONSUMPTION CAPACITY AND COST IN THE APPLICATION OF A GEOTHERMAL ENERGY FACILITY

Following the application of a geothermal energy facility, the energy capacity by facility was estimated based on calculations of geothermal energy production capacity equivalent to 11% of the expected energy consumption capacity of a building structure, with a compulsory supply rate for 2013 of 100% and 25% of geothermal system capacity. The calculation order of geothermal energy installation capacity was as follows: (1) the expected energy consumption capacity of the target building was calculated; (2) the geothermal energy production capacity was estimated by multiplying the compulsory supply rate by the expected energy consumption capacity; (3) the energy production capacity per unit capacity was estimated by multiplying unit energy production capacity by geothermal energy with the application of a modification factor; and (4) installation capacity was calculated by dividing the geothermal energy production capacity by the energy production capacity per unit capacity by geothermal energy. The geothermal energy production capacity is thus a modified value of energy capacity produced annually using a geothermal energy facility, and the expected energy consumption capacity is the expected annual energy consumption capacity.

TABLE 4: Geothermal energy application capacity of cases.

Case 1	Anticipated energy usage	1,271,703 kWh/year
	Geothermal energy production	139,887 kWh/year
	Standard capacity	99.00 kW
	100% installation capacity of standard capacity	105.80 kW
	25% installation capacity of standard capacity	25.65 kW
Case 2	Anticipated energy usage	2,524,596 kWh/year
	Geothermal energy production	277,706 kWh/year
	Standard capacity	194.00 kW
	100% installation capacity of standard capacity	208.38 kW
	25% installation capacity of standard capacity	53.80 kW
Case 3	Anticipated energy usage	8,159,820 kWh/year
	Geothermal energy production	897,580 kWh/year
	Standard capacity	628.00 kW
	100% installation capacity of standard capacity	658.80 kW
	25% installation capacity of standard capacity	167.20 kW

Table 5: Annual total energy demand quantity and use cost of Case 1.

Description		100% application of standard capacity	25% application of standard capacity
Energy usage (kWh/year)	Heating	59,109.30	66,188.76
	Air-conditioning	25,062.46	23,868.33
	Hot-water supply	18,096.71	18,096.71
	Lighting	31,166.15	31,166.15
	Ventilation	31,146.17	31,146.17
	Total	164,580.78	170,466.12
Energy use cost (Won/year)	Heating	5,179,748	5,800,121
	Air-conditioning	2,196,223	2,091,582
	Hot-water supply	1,440,909	1,440,909
	Lighting	2,731,090	2,731,090
	Ventilation	2,729,339	2,729,339
	Total	14,277,308	14,793,041

The expected energy consumption and geothermal energy production capacities were estimated by applying 371.66 (kWh/m^2/year) for unit energy consumption capacity and 1.73 for the modification factor by use. The applied locality coefficients were 0.99 in Case 1 (Gyeonggi region), one in Case 2 (Daejeon region) and 0.98 in Case 3 (Gyeongbuk region). The installation capacity actually selected based on the above process was applied to estimate expected energy consumption capacity, taking the dimensions and formations of each facility into account. Geothermal energy production capacity was calculated by applying 11% of the expected consumption capacity, the compulsory supply rate. The standard capacity represents the capacity of a geothermal energy facility required to satisfy the geothermal energy production capacity. Table 4 shows the application capacity of geothermal energy facilities by region.

Case 1, which appeared to have the best average heat transmission value for the external wall of 0.619 W/m^2 K (K: heat transmission coefficient) in the basic insulation standard, used electricity as the energy source, with the exception of hot-water facilities, which relied on an urban gas boiler.

When applying 100% of the standard capacity, a higher level of cooling energy consumption was achieved compared to the application of 25% of the standard capacity. This may be because the power of the circulation pump for geothermal exchange is greater than that of other cooling load facilities, due to its relatively large size. For the total cooling capacity of the EHP, 191.4 kW was applied, with 3.75 kW for the consumption power of auxiliary devices, 105.8 kW for geothermal heat and 4.4 kW for the circulation pump. Table 5 shows the total annual energy consumption capacity and consumption cost of the geothermal energy facility by region, specifically for Case 1.

The reduction rate of the annual energy consumption capacity and the cost in the case of 100% compulsory installation capacity were shown to be 7.44% and 7.52%, respectively, when compared with cases where no geothermal energy facilities were present. The reduction rate in geothermal heat in the case of 25% compulsory installation capacity was shown to be 50% of the level of 100% compulsory installation capacity, indicating that when applying a geothermal system, differences may arise from different capacities of the geothermal circulation pump, which is known to be the primary cause of power consumption. The circulation pump in the geothermal system is running continuously, and its application capacity is determined depending on geothermal installation capacity. When a large capacity is loaded, the value increases, but the rate of increase is not proportionate to the installation capacity. The selection of an appropriate capacity for the circulation pump of a geothermal system is therefore considered to be a factor that can improve energy efficiency in the application of such a system.

The energy consumption per unit area is used as a criterion for the calculation of total energy consumption, but the gross floor area is used to calculate the lighting and ventilation load, while the cooling and heating areas are used to calculate the hot-water supply load. For this reason, the total energy consumption capacity may differ depending on the ratio of cooling and heating areas over the gross floor area. However, the consumption capacity per unit area shows the energy reduction effect of the applied geothermal energy facility regardless of the entire building structure and can be used as a good criterion for the evaluation of these facilities. Annual energy consumption capacity and cost per unit area are indicated in Table 6. Table 7 provides a comparison of the total annual energy consumption and cost.

TABLE 6: Annual energy demand quantity per unit area and use cost of cases.

Description			100% application of standard capacity	25% application of standard capacity
Case 1	Energy usage	Does not apply	111.42	
		Apply	103.13	107.27
		Reduction capacity	8.29 kWh	4.15 kWh
		Reduction rate (%)	7.44	3.72
	Use cost	Does not apply	9662	
		Apply	8935	9298
		Reduction capacity	727 Won	364 Won
		Reduction rate (%)	7.52	3.77
Case 2	Energy usage	Does not apply	101.28	
		Apply	91.15	99.70
		Reduction capacity	10.13 kWh	1.58 kWh
		Reduction rate (%)	10.00	1.56
	Use cost	Does not apply	10,007	
		Apply	9,027	9,854
		Reduction capacity	980 Won	153 Won
		Reduction rate (%)	9.79	1.53
Case 3	Energy usage	Does not apply	117.23	
		Apply	108.68	113.95
		Reduction capacity	8.55 kWh	3.28 kWh
		Reduction rate (%)	7.29	2.80
	Use cost	Does not apply	11,177	
		Apply	10,349	10,858
		Reduction capacity	828 Won	319 Won
		Reduction rate (%)	7.41	2.85

Table 8 shows annual primary energy consumption per unit area by new renewable energy facility. Geothermal heat ranked higher in the category of annual primary energy consumption per unit area, which resulted from employing kerosene with an energy conversion factor of 1.1 for hot-water supply as the energy source.

TABLE 7: Annual energy demand quantity and use of cases.

Description			100% application of standard capacity	25% application of standard capacity
Case 1	Energy usage	Does not apply	176,366.72	
		Apply	164,580.78	170,466.12
		Reduction capacity	11,785.94 kWh	5,900.60 kWh
		Reduction rate (%)	6.68	3.35
	Use cost	Does not apply	15,310,149	
		Apply	14,277,308	14,793,041
		Reduction capacity	1,032,841 Won	517,108 Won
		Reduction rate (%)	6.75	3.38
Case 2	Energy usage	Does not apply	302,151.68	
		Apply	274,793.40	297,873.27
		Reduction capacity	27,358.28 kWh	4,278.41 kWh
		Reduction rate (%)	9.05	1.42
	Use cost	Does not apply	29,790,302	
		Apply	27,144,843	29,376,666
		Reduction capacity	2,645,459 Won	413,636 Won
		Reduction rate (%)	8.88	1.39
Case 3	Energy usage	Does not apply	1,018,439.93	
		Apply	956,066.67	996,005.34
		Reduction capacity	62,373.26 kWh	22,434.59 kWh
		Reduction rate (%)	6.12	2.20
	Use cost	Does not apply	97,315,248	
		Apply	91,283,753	95,145,822
		Reduction capacity	6,031,495 Won	2,169,426 Won
		Reduction rate (%)	6.20	2.23

The comparison of energy consumption and cost per unit area cannot be perfect, due to differences in equipment characteristics, such as the installation situation of the energy load facility and architectural characteristics, such as window area rate and the insulation performance of walls. However, it can be used as a reference for selecting an appropriate geothermal facility for a building structure, through a comparison of cases of different scales that are characterized as having similar energy consumption at a certain level and which do not require the application of cooling and heating areas. The effect of geothermal heat was shown to be best when applying 100% of the standard capacity. Apart from the installation size, geothermal heat effects are also determined by the energy consumption behaviour of the building structure itself, and the circulation pump of an auxiliary installation is a factor that cannot be ignored. Table 9 shows the energy consumption and cost characteristics for the application of a new renewable energy facility for each of the three cases.

TABLE 8: Annual primary energy demand quantity of cases.

Description		100% application of standard capacity	25% application of standard capacity
Case 1 Energy usage	Does not apply	293.41	
	Apply	270.61	281.98
	Reduction capacity	22.80 kWh	11.43 kWh
	Reduction rate (%)	7.77	3.90
Case 2 Energy usage	Does not apply	244.91	
	Apply	212.60	240.83
	Reduction capacity	32.31 kWh	4.08 kWh
	Reduction rate (%)	13.20	1.67
Case 3 Energy usage	Does not apply	307.98	
	Apply	284.46	298.93
	Reduction capacity	23.52 kWh	9.05 kWh
	Reduction rate (%)	7.64	2.94

When comparing primary energy consumption per unit area, which is the currently used criterion of energy efficiency rating for domestic building structures, a higher saving rate was found in Case 2, with 100% of the standard capacity, compared to Case 3, with 25% of the standard capacity. This is thought to be a result of the basic energy consumption of a building structure, i.e., the decrease in the window area rate of the external wall area resulted in a remarkable decrease in the required heating energy accompanied by the maximization of the effect of the geothermal heating facility and a greater decrease in the power consumption of the EHP system in accordance with reductions in the cooling and heating energy of the geothermal system. Table 10 shows the annual primary energy consumption per unit area for each application of a new renewable energy facility.

TABLE 9: Annual energy demand per unit area of cases.

Description		Case 1	Case 2	Case 3
100% application of standard capacity	Demand quantity reduction (%)	6.68	10.00	7.29
	Use cost reduction (%)	6.75	9.79	7.41
25% application of standard capacity	Demand quantity reduction (%)	3.35	1.56	2.80
	Use cost reduction (%)	3.38	1.53	2.85

TABLE 10: Annual primary energy demand per unit area of cases.

Description		Case 1	Case 2	Case 3
100% application of standard capacity	Demand quantity reduction (%)	7.77	13.20	7.64
25% application of standard capacity	Demand quantity reduction (%)	3.90	1.67	2.94

13.8 LCC ANALYSIS BY GEOTHERMAL ENERGY FACILITY

13.8.1 REPAIR AND REPLACEMENT CYCLE OF THE APPLIED GEOTHERMAL ENERGY FACILITY

In the LCC analysis conducted in this study, geothermal energy repair and replacement were based on an assumption of 40 durable years. To obtain more diverse results depending on the analysis period, the analysis was performed using a 10-year cycle. For items included in the specification, the rate and level of repair and repair and replacement cycles were applied, but when the items were not included in the specification, the durable years of items similarly categorized were applied. The LCC analysis used in this study was not for the entire building structure, but for the geothermal energy facility, and only the components of these facilities were analysed.

When calculating maintenance cost in terms of the repair cycle of the geothermal energy facility by primary process, the rate of repair level and the replacement rate of the geothermal heat pump were 10% at five years and 100% at 10 years. However, a replacement rate of 100% at 11 years was obtained for the expansion tank. When a new renewable energy facility is installed, a structure should be fitted to provide support for the fixation of a buttress and pump. A number of general processes usually applied to general facility work were also included, and the replacement rate of items was likewise analysed. The rate of repair level of the equipment of major components was identical to that of equipment and materials for general equipment work in which a heat pump was used, and there are many components with a short repair cycle; the repair cost was therefore shown to be high.

13.8.2 LCC ESTIMATION OF A GEOTHERMAL ENERGY FACILITY

LCC was calculated using the present value method. In the case of geothermal energy facility equipment for items that only have a replacement cycle, this was converted to nonrecurring cost for every repair cycle, while for items that have periodic repair and replacement cycles, this was converted to nonrecurring cost every repair cycle. Subsequently, all converted values and nonrecurring costs were summed to estimate maintenance cost. The discount rate applied to the estimation of maintenance cost was calculated using a real discount rate of 1.02%, obtained from the inflation and nominal discount rates based on the deposit interest rate for seven years.

The rate of increase of the average electricity fee between 2006 and 2012 was used to establish the inflation rate by referring to Monthly Energy Statistics issued in May, 2013; the value of 1.06% was used as the real discount rate obtained from the calculation of the rate of increase of a nominal discount to estimate energy savings cost. The initial investment cost was estimated by adding the costs of the applied geothermal energy facility calculation specification to variations in the amount accompanied by the replacement and dismantlement of the conventional facility. The construction cost of the wiring system installation for power was excluded, because it is not included in geothermal energy facility construction.

13.8.2.1 CALCULATION OF INITIAL INVESTMENT COST OF A GEOTHERMAL ENERGY FACILITY

To calculate the initial investment cost of a geothermal system, the installed capacity of the existing facility should be changed. In Case 1, when 100% of the standard capacity was applied, a 116 kW outdoor equipment item was removed from the conventional EHP cooling and heating facilities and a new 23 kW outdoor equipment item was installed to preserve capacity. An indoor facility using the geothermal refrigerant method is compatible with the item of EHP outdoor equipment, and there was no influence on the construction cost of indoor facilities. The cost of the 116 kW outdoor equipment item was subtracted, and the cost of the 23 kW outdoor equipment item was added. When 25% of the standard capacity was applied, 75 kW outdoor equipment was removed and replaced with 52.2 kW outdoor equipment. In Case 2, when 100% of the standard capacity was applied, 52.2 kW, 71.8 kW and 78.4 kW EHPs were replaced, leading to a reduction in the initial investment cost. When 25% of the standard capacity was applied, with the substitution effect of a 52.2 kW EHP, the initial investment cost decreased. In Case 3, seven items of 46.4 kW EHP outdoor equipment and four items of 58 kW equipment, together with an item of 75.4 kW equipment were affected by the change. The construction costs of geothermal facilities have risen since the public announcement of the establishment of a base price in 2013. The design cost was calculated based on actual design drawings. In geothermal energy facility construction work, the drilling and installation costs of a geothermal exchanger were shown to be higher than the installation cost of geothermal equipment, along with the installation of a heat pump.

The capacity of the geothermal energy facility can be adjusted to be equivalent to the conventional cooling and heating facility; EHP equipment and installation costs that can be adjusted by case according to the application capacity were included in the initial investment cost of the geothermal system, but their value-added tax was not. The initial investment costs calculated with the construction cost for each geothermal energy type, as calculated based on the design drawings and adjustments in the capacity of the conventional load facilities, are provided in Table 11.

TABLE 11: Initial investment of geothermal energy facilities of cases.

Cases	100% application of standard capacity	25% application of standard capacity
Case 1	177,413,214	57,481,291
Case 2	282,115,135	87,934,371
Case 3	684,896,176	206,499,483

13.8.2.2 CALCULATION OF MAINTENANCE COST OF A GEOTHERMAL ENERGY FACILITY

The maintenance cost was obtained by adding replacement cost to repair cost. For items that only have a replacement cycle, this was converted to a nonrecurring cost for every replacement cycle, while for items that have periodic repair and replacement cycles, this was converted to a nonrecurring cost for every repair cycle. All costs were then added to the nonrecurring cost of the replacement cycle to estimate maintenance cost by analysis period. The maintenance cost of the geothermal system was determined to be higher when 100% of the standard capacity was applied. In the analysis of maintenance cost, it was found that with the exception of drilling work, most of the geothermal facility construction work in the three cases was similar to general cooling and heating construction work where a heat pump is applied; however, the maintenance cost was calculated as high due to a rise in repair and replacement costs resulting from the five-year repair cycle and from a 10-year replacement cycle for the heat pump.

TABLE 12: The initial investment of geothermal energy facilities of cases.

Case	Analysis year	Application (%)	Initial investment	Maintain cost	LCC
Case 1	10	100	177,413,214	45,082,942	222,495,156
	25	57,481,291	22,929,864	80,411,155	
	20	100	177,413,214	114,814,736	292,227,950
	25	57,481,291	52,607,724	110,089,015	
	30	100	177,413,214	170,715,471	348,128,685
	25	57,481,291	78,901,211	136,382,502	
	40	100	177,413,214	182,417,96	359,831,181
	25	57,481,291	86,793,300	144,274,591	
Case 2	10	100	282,115,135	78,591,384	360,706,519
	25	87,934,371	29,083,243	89,533,243	
	20	100	282,115,135	201,614,252	483,729,387
	25	87,934,371	70,780,448	158,714,819	
	30	100	282,115,135	303,539,695	585,654,830
	25	87,934,371	106,299,141	194,233,512	
	40	100	282,115,135	321,713,745	603,828,880
	25	87,934,371	114,652,882	202,587,253	
Case 3	10	100	684,896,176	176,172,862	861,069,038
	25	206,499,483	64,219,177	270,718,660	
	20	100	684,896,176	447,156,081	1,132,052,257
	25	206,499,957	157,211,330	363,710,813	
	30	100	684,896,176	708,458,855	1,393,355,031
	25	206,499,483	236,056,719	442,556,202	
	40	100	684,896,176	743,348,843	1,428,245,019
	25	206,499,483	248,408,375	454,907,858	

13.8.2.3 LCC CALCULATION OF A GEOTHERMAL ENERGY FACILITY

Based on the analysis of the initial investment and the maintenance costs of the geothermal energy facility in each case, the LCC was calculated for every 10-year analysis period. When 100% of the standard capacity was applied in all cases, the LCC of the geothermal energy facility increased as the analysis period increased. The LCC of the geothermal energy facility by analysis period for each case is shown in Table 12.

TABLE 13: Energy savings cost by applying the geothermal energy facilities of cases.

Case	Analysis year	100%	25%
Case 1	10	9,750,941	4,881,961
	20	18,526,071	9,275,367
	30	26,423,043	13,229,110
	40	33,529,739	16,787,188
Case 2	10	24,975,493	3,905,093
	20	47,451,603	7,419,390
	30	67,678,450	10,581,999
	40	85,881,127	13,428,114
Case 3	10	56,942,694	20,481,317
	20	108,186,936	38,912,998
	30	154,302,990	55,500,157
	40	195,804,051	70,427,381

13.8.2.4 A COMPARATIVE ANALYSIS OF THE LCC OF A GEOTHERMAL ENERGY FACILITY

In terms of geothermal energy, the cost savings effect did not appear to be in proportion to the application capacity, but differed in each case. Table 13 specifies the energy savings cost that accompanied the application of a geothermal energy facility in each case. The analysis showed that it is difficult to expect economic benefits from the application of a geothermal energy facility. Even though with the geothermal system the initial investment cost can be compensated for by adjusting the energy load capacity for the facility, this was not economically feasible, because of high initial investment and maintenance costs and low energy savings costs. This implies that the circulation pump required to operate the geothermal system consumes significant electricity, thus minimizing the energy-saving effect. Its initial investment cost is also high. To improve the economic feasibility of the geothermal energy system, less power-intensive circulation pumps and devices should therefore be selected and the initial investment cost should be reduced.

It was not effective to estimate the period required to recoup the initial investment with the analysed data, since the difference between energy savings and initial investment costs was high. The return rate of LCC input through energy savings cost was thus used. In Case 1, the return rate of LCC for a new renewable energy facility increased as the analysis period increased, which indicates that the increase in energy savings cost was greater than that in the LCC of a geothermal facility. Thus, if the initial investment and maintenance costs of the geothermal energy facility are improved, the economic effect will be greater. The LCC analysis of the geothermal energy facility applied to each case is provided in Table 14.

13.9 CONCLUSIONS

This study aimed to analyse energy consumption in the application of new renewable energy systems to a public office building based on compulsory application criteria. The study also sought to determine an effective plan for the selection of a new renewable energy facility that takes economic feasibility into account by performing a comparative analysis of LCCs of new renewable energy facilities and energy savings costs from their application. Two main research findings were obtained. First, energy consumption and costs related to the application of a geothermal energy facility were analysed in three cases, to arrive at effective data for the selection of geothermal energy facility types. Second, the energy cost reduction effect relating to the application of a geothermal energy facility was examined, and an effective plan for the selection of a geothermal energy facility was presented by calculating the LCC using the initial investment and maintenance costs of a geothermal energy facility, the latter obtained by applying repair and replacement rates based on the construction specifications of each facility. It is believed that the results of this study can be utilized as an effective plan for the selection of a geothermal energy facility based on economic feasibility. In addition, the energy-saving effects on energy consumption and primary energy consumption will also be utilized as fundamental data in the understanding and selection of geothermal energy facilities.

TABLE 14: Accumulated operation and maintenance cost (LCCO2 cost excluded).

Case	Analysis year	Description	100%	25%
Case 1	10	LCC	222,496,156	80,411,155
		Energy savings cost	9,750,941	4,881,961
		LCC collection ratio	4.38%	6.07%
	20	LCC	292,227,950	110,089,015
		Energy savings cost	18,526,071	9,275,367
		LCC collection ratio	6.36%	8.43%
	30	LCC	348,128,685	136,382,502
		Energy savings cost	26,423,043	13,229,110
		LCC collection ratio	7.59%	9.70%
	40	LCC	359,831,181	144,274,591
		Energy savings cost	33,529,739	16,787,188
		LCC collection ratio	9.32%	11.64%
Case 2	10	LCC	360,706,519	428,982,726
		Energy savings cost	24,975,493	3,905,093
		LCC collection ratio	6.92%	3.32%
	20	LCC	483,729,387	158,714,819
		Energy savings cost	47,451,603	7,419,390
		LCC collection ratio	9.81%	4.67%
	30	LCC	585,654,830	194,233,512
		Energy savings cost	67,678,450	10,581,999
		LCC collection ratio	11.56%	5.45%
	40	LCC	603,828,880	202,587,253
		Energy savings cost	85,881,127	13,428,114
		LCC collection ratio	14.22%	6.63%
Case 3	10	LCC	861,069,038	270,718,660
		Energy savings cost	56,942,694	20,481,317
		LCC collection ratio	6.61%	7.57%
	20	LCC	1,132,052,257	363,710,813
		Energy savings cost	56,942,694	20,481,317
		LCC collection ratio	6.61%	7.57%
	30	LCC	1,393,355,031	442,556,202
		Energy savings cost	154,302,990	55,500,157
		LCC collection ratio	11.07%	12.54%
	40	LCC	1,428,245,019	454,907,858
		Energy savings cost	195,804,051	70,427,381
		LCC collection ratio	13.71%	15.48%

However, the capacity of a new renewable energy facility was not simply calculated based on its energy production capacity; to analyse energy efficiency and economic feasibility in a more practical way, the energy production capacity and use efficiency of each new renewable energy facility should be utilized. The energy production capacity should be calculated based on the characteristics of each facility, and the energy-saving capacity resulting from the application should also be considered. Future work should analyse the energy use and cost characteristics of building constructions with new renewable energy facilities, and the results should be compared with the energy use characteristics of buildings, evaluated based on design drawings, to secure reliability. A study should also be conducted to prepare selection criteria for an appropriate facility by use, size and characteristics based on the actual performance of each new renewable energy facility.

REFERENCES

1. Furundzic, A.K.; Kosoric, V.; Golic, K. Potential for reduction of CO2 emissions by integration of solar water heating systems on student dormitories through building refurbishment. Sustain. Cities Soc. 2012, 2, 50–62.
2. Korea Energy Economics Institute. Energy information statistics centre. 2012–2017 forecasting energy demand. Available online: http://www.keei.re.kr (accessed on 5 December 2013).
3. Song, J.Y.; Jung, J.W.; Hong, W.H.; Park, H.S. Analysis on the building energy efficiency rating by new renewable energy rating system. Proc. J. Ecol. Environ. 2011, 21, 63–66.
4. Rezaie, B.; Esmailzadeh, E.; Dincer, I. Renewable energy options for buildings: Case studies. Energy Build. 2011, 43, 56–65.
5. Visa, I.; Moldovan, M.D.; Comsit, M.; Duta, A. Improving the renewable energy mix in a building toward the nearly zero energy status. Energy Build. 2014, 68, 72–78.
6. Cucchiella, F.; D'Adamo, I.; Gastaldi, M.; Koh, S.C.L. Renewable energy options for building: Performance evaluations of integrated photovoltaic systems. Energy Build. 2012, 55, 208–217.
7. Francisco, S.S.; Batlles, J. Renewable energy solutions for building cooling, heating and power system installed in an institutional building: Case study in South Spain. Renew. Sustain. Energy Rev. 2013, 26, 147–168.
8. Jung, M.H.; Park, J.C.; Lee, E.G. A study on the application of renewable energy systems to apartment houses. Proc. Architect. Inst. Korea 2009, 28, 591–594.

9. Kang, S.H.; Ryu, S.W.; Hwang, J.H.; Cho, Y.H. An application analysis of renewable energy for public building and an analysis of building energy substitution rate. Proc. Korea Sol. Energy Soc. 2011, 31, 348–353.
10. Kim, M.R.; Lee, K.J.; Park, H.S. A study on the cases of new renewable energy applied buildings in Korea and Germany—focused on solar and geothermal energy cases. Archit. Inst. Korea 2012, 28, 29–37.
11. Yoon, D.I.; Ko, M.J.; Cho, Y.H.; Cho, J.H.; Jang, J.D.; Kim, Y.S. A study on the current status and feasibility of new & renewable energy system with survey. Soc. Living Environ. Syst. Korea 2013, 20, 225–232.
12. Seo, S.H.; Hong, J.H.; Lee, Y.H.; Cho, Y.H.; Hwang, J.H. A study on the analysis of the efficiency of new and renewable energy applied to complex government office buildings. Proc. Korea Sol. Energy Soc. 2013, 33, 356–361.
13. Kim, J.M.; Kim, K.Y. A study on econoic analysis of new renewable energy power (photovoltaic, wind power, small hydro, biogas). Korea Sol. Energy Soc. 2008, 28, 70–77.
14. Kim, H.G.; An, G.H.; Choi, Y.S. LCC analysis for optimized application of renewable energy of eco-friendly school. Archit. Inst. Korea 2011, 27, 83–90.
15. Lee, Y.H.; Seo, S.H.; Kim, H.J.; Cho, Y.H.; Hwang, J.H. Analysis of new & renewable energy application and energy consumption in public buildings. Korean Sol. Energy Soc. 2012, 32, 153–161.
16. Lee, T.J. Development and application of geothermal energy. KSGEE 2009, 4, 6–14.
17. Na, S.M. Current state of world geothermal development and EGS geothermal development. KSGEE 2008, 4, 6–14.
18. Park, J.K. A study on field construction and improvement for high efficiency geothermal system. Ph.D. Thesis, Gyeongsang National University, Jinju, Republic of Korea, 2011.

Author Notes

CHAPTER 1

Conflict of Interest

The authors declare that there is no conflict of interests regarding the publication of this paper.

CHAPTER 2

Acknowledgments

We thank the National Center for Atmospheric Research (NCAR) for making the code and driving data of the CCM3 model available to us. This research was supported by the US Department of Energy (DOE-BER and DOE-NREL), and the Corporate, Governmental, and Foundation sponsors of the MIT Joint Program on the Science and Policy of Global Change.

CHAPTER 3

Conflict of Interest

The authors declare that there is no conflict of interests regarding the publication of this paper.

CHAPTER 4

Conflict of Interest

The authors declare that there is no conflict of interests regarding the publication of this paper.

Acknowledgments

The authors acknowledge the E-OBS dataset from the EU-FP6 project ENSEMBLES (http://ensembles-eu.metoffice.com) and the data providers in the ECA&D project (http://www.ecad.eu). The ENSEMBLES (http://ensemblesrt3.dmi.dk/) database provided the RCM data used in this study. The SoDa database (SoDa Contract DG "INFSO" IST-1999-12245) provided the observed irradiance data used. This research work was funded by the EU FP7 ECLISE project (Grant Agreement no. 265240).

CHAPTER 5

Acknowledgments

The authors would like to thank the Norwegian Research Council through Norwegian University of Science and Technology for the financial support. The authors would like to also thank P.C.D. Milly for data on the future global runoff projections. The rest of the data sources mentioned under data section are also acknowledged for the data access and use.

CHAPTER 6

Acknowledgments

This study was financially supported by the European Commission through the FP7 ECLISE project to Wageningen University.

CHAPTER 7

Acknowledgments

The authors thank the German Federal Ministry of Education and Research (bmb+f) and the Free State of Bavaria for funding the project GLOWA-Danube. The provision of meteorological and hydrological data by the German Weather Service (DWD), the Austrian Weather Service (ZAMG) and the Environmental Agency of the Free State of Bavaria (LfU) is gratefully acknowledged.

CHAPTER 8

Conflict of Interest

The author declares no conflict of interest.

CHAPTER 9

Conflict of Interest
The authors certify that there is no conflict of interests with any financial organization regarding the material discussed in the paper.

CHAPTER 10

Acknowledgments
The authors kindly acknowledge the financial support provided by the Natural Sciences and Engineering Research Council of Canada.

Conflict of Interest
The authors declare no conflict of interest.

CHAPTER 11

Acknowledgments
This paper is financially sponsored by the National Science Council under Grants NSC-101-2622-E-027-025-CC3, NSC-101-2219-E-027-006, and NSC-99-2221-E-027-069-MY3.

CHAPTER 12

Acknowledgments
The current speed and solar irradiation data on the SHELL Sabah Water Platform are provided by SHELL Sarawak Sdn. Bhd. and dated back in year 2008.

CHAPTER 13

Author Contributions
In this paper, Sangyong Kim developed the research ideas and organized research flow; Young Jun Jang implemented research program and collect-

ed data; Yoonseok Shin participated analysis of case study part; Gwang-Hee Kim completed the writing work of corresponding parts.

Conflict of Interest

The authors declare no conflict of interest.

Index

For Product Safety Concerns and Information please contact our EU
representative GPSR@taylorandfrancis.com
Taylor & Francis Verlag GmbH, Kaufingerstraße 24, 80331 München, Germany

www.ingramcontent.com/pod-product-compliance
Ingram Content Group UK Ltd.
Pitfield, Milton Keynes, MK11 3LW, UK
UKHW022159190926
475702UK00003B/111
* 9 7 8 1 7 7 1 8 8 4 3 1 0 *